THE PLANT HUNTERS

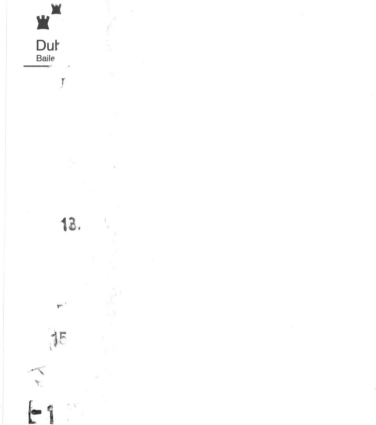

THE PLANT HUNTERS

Two hundred years of adventure and discovery around the world

**TOBY MUSGRAVE, CHRIS GARDNER
AND WILL MUSGRAVE**

First published in the United Kingdom in 1998 by
Ward Lock

This paperback edition first published in 1999 by
Seven Dials, Cassell & Co.
The Orion Publishing Group
Wellington House, 125 Strand
London, WC2R 0BB

Distributed in the United States of America by
Sterling Publishing Co., Inc.
387 Park Avenue South,
New York, NY 10016-8810

A CIP catalogue record for this book is available
from the British Library

ISBN 1 84188 001 9

Designed by Yvonne Dedman
Maps by ML Design
Colour separations by Reed Digital, Ipswich, England
Printed in Hong Kong

Cover Illustrations:
Front (*from top, left to right*): *Banksia Integrifolia*, from Banks's
Florilegium (National History Museum, London); Joseph Banks, aged
thirty, painted by Joshua Reynolds (courtesy of The National Portrait
Gallery, London); David Douglas, attributed to Sir Daniel McNee,
(by permission of The Linnean Society of London); *Camellia
saluensis* (John Glover); *Zantedeschia aethiopica,* (Clive Nichols);
Rhododendron thomsonii by Walter Fitch from *Rhododendrons of the
Sikkim-Himalaya* by Sir Joseph Dalton Hooker, 1849, (Linnean
Society of London / Bridgeman Art Library, London)
Back *(from top): Rhododendron thomsonii* by Walter Fitch from
Rhododendrons of the Sikkim-Himalaya by Sir Joseph Dalton Hooker,
1849 (Linnean Society of London / Bridgeman Art Library, London);
Banksia Integrifolia, from Banks's *Florilegium* (National History
Museum, London)

CONTENTS

Authors' Acknowledgements 7

Introduction 9

1. **Endeavour and Expansion:** Sir Joseph Banks 13

2. **To the Fairest Cape:** Francis Masson 39

3. **A Walk on the Wild Side:** David Douglas 55

4. **On Top of the World:** Sir Joseph Dalton Hooker 79

5. **Fortune Favours the Brave:** Robert Fortune 105

6. **Brothers in the Nursery:** The Lobbs and the Veitch Dynasty . . . 131

7. **Chinese Puzzle:** Ernest Wilson 155

8. **A Forrest of Rhododendrons:** George Forrest 177

9. **Kingdom of the Blue Poppy:** Frank Kingdon-Ward 199

Bibliography 218

Illustration Acknowledgements 220

Index 221

AUTHORS' ACKNOWLEDGEMENTS

We would like to express our gratitude to the following people, without whom this book would have remained a collection of muddled ideas: Jan Clamp for steering us on to the right path and, without fail, offering wise counsel; Wanda, Chris, Elena and Jonathan Millis for their wonderful and frequent hospitality and their never-ceasing good humour; our agent Fiona Lindsay for all her hard work and patience in dealing with us.

We should also like to thank all those who have assisted in the transformation from manuscript to book; in particular, Annie Lee, the copy-editor, and, at Cassell, Barry Holmes, Jane Birch and Clare Churly.

Without a doubt, the most heartfelt 'thank you' must go to our long-suffering but extremely tolerant and supportive families and friends and, of course, to Terry the cat.

Good God. When I consider the melancholy fate
of so many of botany's votaries, I am tempted to ask whether
men are in their right mind who so desperately risk life
and everything else through the love of collecting plants.

Carl Linnaeus, *Glory of the Scientist* (1737)

INTRODUCTION

HUNDREDS OF THOUSANDS OF US ENJOY GARDENING AND VISITING famous parks and gardens. Yet few of us, as we admire the beautiful and diverse range of plants around us, stop to wonder where they come from, and fewer still think about how these plants came to be here in the first place.

How many of us, for example, know that the explorer who found over 300 rhododendron species was one of two survivors of a party attacked in a rebellious uprising and had an escape worthy of a member of the Special Forces; that the man responsible for establishing the tea industry in India single-handedly fought a gun battle with pirates while running a high fever; that the plant hunter who introduced many conifers to our landscape was gored to death by a bull; or that the discovery of the Himalayan rhododendrons resulted in a kingdom being annexed into the British Empire?

Plants have been gathered since the dawn of time. The first record of plant hunting dates from 1495 BC and tells of an expedition sent by Queen Hatshepsut from Egypt to Somalia to collect incense trees (*Commiphora myrrha*). The Romans took many of their plants with them as their Empire expanded, and, later on, medieval monasteries exchanged plants across Europe. The Romans and the monks introduced new plants to a number of countries, but there is little evidence that they engaged in active plant hunting.

As peaceful civilizations flourished, so did the garden. Growing newly discovered plants became a source of pleasure and excitement. Up until the mid-sixteenth century most new plants arriving in Britain came from Europe, but for the next eighty or so years the Middle East became the source of horticultural novelty. From 1620 onwards, as North America was increasingly explored and colonized, a wealth of species was sent back to Britain from Canada and Virginia. The most celebrated plant hunters of the seventeenth century were John Tradescant the elder (*c.* 1570–1638) and his son, John Tradescant the younger (1608–62). John the elder made trips to the Low Countries and in 1618 visited Russia, but it was through his membership of the Virginia Company that he received many first introductions of North American species, including the stag's horn sumach (*Rhus typhina*) and the spiderwort (*Tradescantia virginiana*). John the younger made three trips to Virginia, in 1637, 1642 and 1654. Among the species he is credited with introducing are the swamp cypress (*Taxodium distichum*), the tulip tree (*Liriodendron tulipifera*) and the American cowslip (*Dodecatheon meadia*). The work of the Tradescants and others established the role of the plant hunter and stimulated the gardener's passion for creating plant collections, which continues to this day. Nevertheless, on occasion enthusiasm got the better of good judgement. In 1635 a house in Hoorn, in the Netherlands, was exchanged for three tulip bulbs, and in the late nineteenth century fanatical orchid growers were prepared to pay over 1,000 guineas for a new plant.

The last 200 years have seen enormous changes in garden design. Nothing has transformed the look of the garden more than exotic foreign plants. The thousands of plants such as conifers, rhododendrons, magnolias, camellias, jasmines, clematis and lilies that are commonly grown in gardens around the world stand as testament to the bravery and determination of an eclectic group of botanists, explorers and collectors. However, their work would not have been possible without the support of a range of different sponsors. The Royal Horticultural Society, founded in 1804 with the express purpose of 'the improvement of horticulture' – a philosophy pursued with equal vigour today – funded the travels of a number of important collectors, including David Douglas, Robert Fortune, George Forrest and Frank Kingdon-Ward. By the mid-nineteenth century, commercial nurseries, foremost among them the Veitch Nursery, had come to recognize the profits that could be made from plant hunting. The many new species brought back by their directly funded agents not only produced huge commercial rewards but also further enriched the diversity of the garden – a benefit that continues today, long after those pioneering nurseries are gone.

The scientific and economic value of plant hunting has been much over-looked in recent years. The Royal Botanic Gardens at Kew sent numerous plant hunters to all parts of the globe and invested heavily in the study of new introductions. Kew also instigated the policy of transferring economically important plants between countries. This resulted in the establishment of plantations in many British colonies, and the wealth created by trade in rubber, cinchona (quinine), tea and other economic crops played a vital role in imperial expansion.

This book tells the stories of the plant hunters and how their adventures and discoveries changed the garden landscape for ever. However, before we begin, a distinction needs to be made between 'plant discovery' and 'plant introduction'. In this book plant discovery is taken to mean the first time a new plant was recorded by science. This often took the form of dried and pressed parts of the plant (a herbarium specimen) being sent to a botanical establishment such as Kew. Here it would be examined, classified, named and added to the roll of new plant discoveries. Plant introduction, on the other hand, is defined as the first time that living plant matter – seed, cutting or whole plant – was brought back to Britain. For example, *Davidia involucrata*, the handkerchief tree, was discovered by Père David in 1869 but introduced to Britain by Ernest Wilson in 1901. Often the two happened simultaneously – for example, Sir Joseph Hooker found and introduced his Himalayan rhododendrons between 1849 and 1851.

The plant hunters were exceptional men who, for a variety of reasons, dedicated their lives to increasing the understanding of botany and horticulture. Between them, they collected tens of thousands of new species, and gardeners all around the world owe them an enormous debt. Their heroic endeavours over the past 200 years were as influential as the garden designers and proselytizers in determining the course of garden-making. Fashions in garden design come and go, but they leave their mark and act as a reminder of the work of these men.

There are no longer rich patrons to sponsor trips lasting several years, yet plant hunting still continues. Today's adventurer is equipped with modern technology and transport and with new horticultural techniques, which mean that the discoveries they make have a much better chance of survival. However, contemporary explorers who take up the challenge of inhospitable terrain and protracted political negotiations are no less intrepid than their predecessors; and as long as there remain isolated valleys in places such as South America, Tibet and Papua New Guinea waiting to be explored, there will be adventurers to take up the challenge.

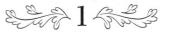

1

ENDEAVOUR AND EXPANSION

Sir Joseph Banks

(1743–1820)

IN THIS TALE OF UNSUNG HEROES ONE MAN STANDS ABOVE ALL OTHERS AS the father of modern plant hunting – Sir Joseph Banks. Even in a century as eclectic as the eighteenth, with its full complement of eccentrics and geniuses, there was no one quite like Sir Joseph. At first glance his story appears to be that of a rich boy with sufficient private means to indulge an unorthodox lifestyle, but this is far from a fair assessment of this complex character. Even though he was not an outstanding scholar, Banks was probably the most influential scientist of his time. His lifelong passions for botany and discovery had ramifications that he could not have foreseen and that are still felt today.

Left: Luxuriant tropical vegetation, warm, clear seas and friendly people welcomed the crew of the *Endeavour* to Tahiti in April 1769.

Above: Joseph Banks: the self-assured and ambitious scientist, shortly after his circumnavigation of the globe with Captain Cook.

The Banks family were landed gentry from Lincolnshire, where their ancestral home was Revesby Abbey. Joseph was born on 2 February 1743 in Argyll Street, London, at a time when the aristocracy dominated the political, social and economic life of Britain. He was born into a family with money and privilege – both of which he was to use wisely throughout his life, even if on occasions he did display the arrogance so often associated with men of his background. He seems to have had a happy childhood and followed the traditional path of the elder son – public school and Oxford. At the age of thirteen, while he was at Eton, Joseph discovered his calling in life. Walking back alone one summer's evening after swimming in the Thames, he was suddenly transfixed by the natural beauty and variety of a wild hedgerow softly illuminated by the evening sunlight. He was captivated by the simple flowers and decided he wanted to know all about them. Showing his characteristic determination and charm, he persuaded the local women who collected herbs for the pharmacists to teach him all they knew. With the aid of a battered copy of Gerard's *Herball* (the standard botanical text of the day) stolen from his mother, he soon became more knowledgeable than his tutors.

Although not highly gifted at school (there are signs that he suffered from dyslexia), Banks took his thirst for botanical knowledge to Oxford University, entering Christ Church in 1760. Here, however, he was forced to use his initiative and family wealth to pursue his studies. At that time the university was going through a period of torpitude, and his Professor of Botany, Humphrey Sibthorp, was not unusual in giving no lectures. On the requested advice of the Cambridge Professor of Botany, John Martyn, Banks 'imported' Israel Lyons to instruct him and his fellow students. He went down from Oxford with an honours degree in December 1763 and went to stay with his mother, who had moved to Chelsea following the death of her husband in 1761. Banks came of age and into his inheritance in 1764, immediately becoming one of the richest young men in Britain: the 9,382-acre Revesby estate yielded over £5,000 per annum and in addition he had income from mining interests in other parts of the country.

By the middle of the eighteenth century it was expected that as part of his education a young man of means would take a Grand Tour to Italy, to see the works of the Renaissance authors, painters, sculptors, architects and gardeners. Works of art were avidly collected and brought back to England, where they helped to stimulate the arts, including the English Landscape School of garden design, which reached its zenith in the works of Lancelot 'Capability' Brown.

Banks, however, said somewhat arrogantly, albeit prophetically: 'Every block-head does that. My Grand Tour shall be one around the globe.' To the horror of his family and friends he secured the position of naturalist aboard the Fisheries Protection vessel HMS *Niger* on her seven-month survey of the remote Labrador and Newfoundland coastline in 1766. His collection of plant species from the trip formed the basis of his herbarium (a collection of dried, pressed and named plant specimens), which in later life was to attain international importance. The collection is now in the British Museum of Natural History, and to look at the dried plant specimens in their original mahogany boxes is to be transported back some 200 years. It is easy to imagine a studious young man working in a poorly lit cabin aboard a rolling, cramped and creaking wooden sailing-boat, carefully examining, recording and preserving his latest treasures.

Two years after his return Banks heard about an expedition that caused him great excitement. The Admiralty and the Royal Society were to send a ship, under Captain James Cook, to the South Seas in order to observe the transit of Venus, an event of astronomical significance. It was believed that this would lead to improved navigational techniques at sea, something the government saw as vital as it started to expand its colonies and began to empire-build. The 1763 peace treaty signed with France at the end of the Seven Years' War had yielded territories in Canada and the West Indies, and the British East India Company now dominated trade in India; the Navy needed to be able to move swiftly and accurately around the world's oceans in order to safeguard Britain's territories and trade routes, particularly against the avaricious activities of other colonial powers such as Holland, Portugal and Spain. This voyage would also give Britain the excuse to monitor her rivals in their quest for colonies and would offer the perfect opportunity to do some exploring. There was a strongly held belief that a southern continent lay waiting to be discovered, and King George III himself gave Captain Cook secret orders to seek out this fabled land of Terra Australis.

Excited by the possibility of collecting flora and fauna from the various stopping points on the voyage, Banks paid the enormous sum of £10,000 for himself and his team of nine men to join the crew. Foremost in this group was Dr Daniel Carl Solander, a much respected Swedish naturalist and a pupil of the great Linnaeus. He was aided by Herman Didrich Sporing, who acted as a secretary of sorts, the scientific draughtsman Sydney Parkinson, and Alexander Buchan and John Reynolds, both charged with drawing landscapes and figures. The remaining four men – two of them boys from the Revesby estate – acted as servants.

The choices of Captain Cook as expedition leader and of the vessel itself, a Whitby collier renamed the *Endeavour*, were at first sight unusual ones but they turned out to be inspired. Cook came from humble origins and had worked his way up through the ranks of the merchant navy before joining the low ranks of the Royal Navy. His able seamanship ensured his promotion – he had particularly impressed the Admiralty with his marine survey of Newfoundland. Cook had the innate instincts of a geographer and a practicality of mind that would enable him to cope in moments of crisis. Despite having very disparate upbringings, Cook and Banks shared a natural sense of authority, were men of common sense and were both at least partially self-educated. Their compatibility can be evinced by the fact that the rather headstrong young Banks and the more cautious Cook, who was fifteen years his senior, do not appear to have had more than one or two minor differences of opinion throughout the whole of the three-year trip.

Cook had chosen the collier, or 'cat', for two simple reasons: he was well acquainted with this type of vessel, and it was flat-bottomed and thus less likely to run on to coral reefs and other hazards than its deeper hulled counterparts. Unfortunately, the *Endeavour*'s virtues of reliability and safety were somewhat counterbalanced by her small size. The ninety-four members of the crew, which included a troop of marines, had to cram themselves and all their provisions on to a craft that weighed just 368 tons and measured only 106 feet long and 29 feet 3 inches at her widest point.

On the afternoon tide of 25 August 1768 and with a fair wind blowing, the *Endeavour* slowly headed out into the English Channel and set a south-westerly course towards the Bay of Biscay. Banks's jubilation at finally being under way was marred two days later when the weather turned foul and to his chagrin the sea-sickness from which he had suffered on the trip to Newfoundland returned. Nevertheless, during the passage to Madeira he and Solander were able to observe and shoot passing wildlife, including porpoises and common storm petrels. The six-day stop-over at Madeira, from 12–18 September, proved to be a disappointment: the season was wrong for any serious plant collecting, and Banks was hampered by the Governor, who insisted on visiting, causing them to lose a precious day's work. Once under way again, Banks and Solander continued their scientific observations, and on 29 September Banks enjoyed the first of many strange dishes when a young shark was caught, gutted

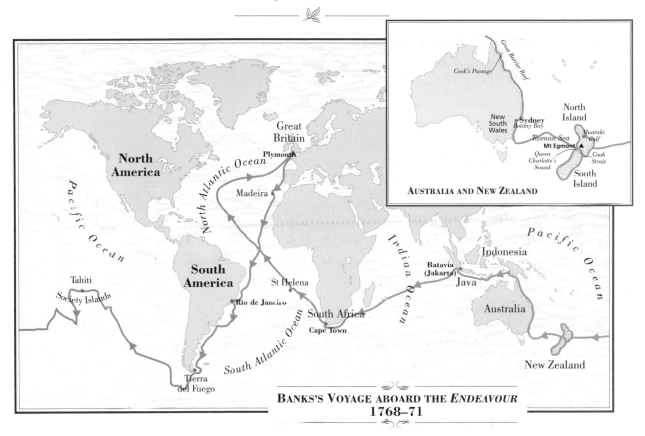

**BANKS'S VOYAGE ABOARD THE *ENDEAVOUR*
1768–71**

and stewed for dinner. (Other exotic foods that he experimented with over the next three years included albatross, dog, rat and kangaroo.)

The steady voyage south was enlivened on 25 October when the *Endeavour* crossed the Equator. The weather was gradually becoming more humid and uncomfortable, and it was with some relief that they reached Rio de Janeiro on 13 November 1768. Cook had assumed that the friendly relationship between Britain and Portugal would ensure a warm welcome, but the Viceroy, Don Antonio de Moura, proved to be exceedingly unfriendly. He made it clear that the *Endeavour* was not wanted and forbade Banks and Solander to go ashore on plant hunting trips. The infuriated Banks was reduced to rummaging through the grasses collected for the ship's livestock, sending his servants on expeditions, and bribing local people to bring plants on board under the pretext of their being 'greens and salading for our table'. In typically pugnacious style he also made a clandestine excursion ashore with Solander. Sneaking from the *Endeavour* before daybreak, they spent a successful day observing and collecting plants before returning under cover of night.

Banks's efforts in Rio resulted in his gathering a total of 316 types of plant, including *Passiflora* species (passion flowers). After a series of further encounters with the Viceroy and twice coming under cannon fire, the *Endeavour* eased her way down the South American coast in early December. As they neared the notorious Cape Horn, Tierra del Fuego was sighted on 11 January 1769 and four days later a suitable anchorage was found in the Bay of Good Success. After dinner some of the crew went ashore and experienced their first contact with an aboriginal people. A large group of Indians appeared on the beach but as the landing party advanced towards them they retreated:

> Dr Solander and myself then walkd forward 100 yards before the rest and two of the Indians advanc'd also and set themselves down about 50 yards from their companions. As soon as we came up they rose and each of them threw a stick he had in his hand away from him and us, a token no doubt of peace, they then walkd briskly towards the other party and wavd to us to follow, which we did and were receivd with many uncouth signs of friendship. We distributed among them a number of Beads and ribbands which we had brought ashore for that purpose at which they seem'd mightly pleas'd.

One of Banks's most useful skills on the *Endeavour*, and indeed one of his most endearing characteristics, was the way in which he was able to defuse potentially dangerous situations and win the trust of strangers. Despite his social status he never patronized the indigenous populations he encountered. Following this initial contact in Tierra del Fuego, three natives were taken back to the *Endeavour* and given a supper of bread and beef. Banks was surprised at the offhand manner with which they treated their alien surroundings and the indifference shown by the rest of the tribe when they were returned ashore.

The next day dawned sunny and warm and, with the prospect of some rewarding plant collecting, an expedition was arranged to explore inland. Initially the going was easy, but soon the party reached a large expanse of waist-high scrub. They struggled on through the scrub as the afternoon passed, but just when matters seemed to be rectifying themselves, Buchan, the artist, was seized with an epileptic fit. A fire was quickly built for him and Banks decided to push on, leaving him in the care of those of the party too exhausted to continue. Before long the weather changed and bitingly cold snow flurries overtook them. Realizing that they could not now return to the safety of the ship, Banks arranged a rendezvous spot in the nearby woods, where it was hoped the whole party would find shelter for the night. Meanwhile he returned to Buchan and was relieved to find him much better. During the trek to the woods, however, first Solander and then Richmond, one of the servants, collapsed from the severe cold. Banks sent four men ahead to build a fire while he attempted to coax the men on, but both were now adamant that they could go no further. When a member of the forward party returned to say that a fire had been started just a quarter-of-a-mile ahead, Solander managed to rouse himself sufficiently to continue. Although he was told that he would die if he remained where he was, Richmond could not move, so Banks had to leave him with a second servant, Dorlton, and a sailor while he accompanied Solander to the camp.

The storm blew hard all night, and dawn brought no respite. Eventually the scattered party was able to regroup, but both Richmond and Dorlton had succumbed to hypothermia. With no provisions, the party were forced to eat a raw vulture which had been shot the previous day and soon after ten o'clock a sombre group of men set out for the *Endeavour*, arriving three hours later, bedraggled and dispirited but with no further casualties.

Banks was finally able to muster an impressive collection of 125 new plants, including *Gaultheria mucronata*, before the *Endeavour* set sail. Following the trials that had beset them in Tierra del Fuego it must have been a great relief to sail calmly around Cape Horn and into the Pacific Ocean. The *Endeavour* made swift progress, and on 13 April 1769 she drifted into Matavia Bay, Tahiti. The stunning scenery – luscious green hillsides, cascading coconut and palm groves, blindingly white sandy beaches and calm azure sea –

Glossy, deep pink fruits festoon the spiky-leafed *Gaultheria mucronata* in autumn. This showy evergreen was collected by Banks from the harsh coast of Tierra del Fuego.

was a paradise compared to the inhospitable land they had visited just a couple of months before. In contrast to the Portuguese in Rio, the Tahitians gave them a much warmer welcome, paddling out to greet their guests in a flotilla of canoes and happily exchanging coconuts, breadfruit, small fish and apples for beads. When the crew of the *Endeavour* made landfall, the locals offered green boughs as a further sign of friendship and walked with the crew for '4 or 5 miles under groves of Cocoa nut and bread fruit trees loaded with a profusion of fruit and giving the most gratefull shade I have ever experienced, under these were the habitations of the people most them without walls: in short the scene we saw was the truest picture of an arcadia of which we were going to be kings that the imagination can form.'

The next day the dignitaries of the tribe, who had previously stayed out of sight, arrived in canoes and motioned for Banks and his companions to accompany them. They were taken to a traditional longhouse, where an old chief presented them with gifts of a cock and hen and long strips of bark. In return Banks gave the chief a large piece of lace and a small handkerchief, with which the chief proudly proceeded to adorn himself. Following this ceremony the party was able to walk freely among the surrounding houses accompanied by attentive female companions. Banks was rather taken aback by their solicitude: 'but as there were no places of retirement, the houses being intirely without walls, we had not an opportunity of putting their politeness to every test that maybe some of us would not have faild to have done had circumstances been more favourable.'

Gradually it became clear that Banks's rapport with the Tahitians and his skills at diplomacy made him the obvious choice as intermediary between the two cultures. This unfortunately meant that he had little time for plant-collecting (although he did find gardenias and a jasmine) because he frequently found himself resolving disputes between the Tahitians and the crew and trying to retrieve goods stolen from the latter by the former. He did, however, have an excellent opportunity to make an ethnological study of the Tahitians, which he set about with his usual enthusiasm. On one occasion a particular subject of Banks's research came a little too close for his comfort. When Banks met Queen Purea, a large woman in her forties who was famous for her nymphomania, he was ushered under her canoe awning into her sleeping area. Here he was somewhat taken aback to find her in the arms of her young consort, Obadee. Purea at once dismissed Obadee, making it plain that she wanted to capture Banks in a similar clinch. Although there is no evidence that she succeeded, Banks was to suffer severe lampooning by his detractors when he returned to Britain.

On 16 April tragedy struck the expedition when Buchan was struck down by another epileptic fit. This time it was more serious, and the following day he died. His death was a severe blow to Banks, for not only had he lost an 'ingenious and good young man' but he now had no way of recording the scenes of everyday life. Parkinson and Reynolds had their own work to carry out, and Banks's artistic skills could be described at best as childlike.

Cook successfully observed the transit of Venus, and with the main task completed there was still time for a circumnavigation of the island and for further studies of the local inhabitants, during which Banks witnessed the art of tattooing (on a young girl's bottom), sampled gastronomic delicacies including vegetable-fed dog and raw fish, and was given the honour of participating in a funeral. The time now came, though, for Cook to act on the King's secret orders and on 13 July, amid much weeping and wailing from 'our friends', Cook and his reluctant crew raised anchor. Before they could set sail, however, the chief priest of the island, Tupaia, demanded that he be taken with them. Cook refused, but Banks felt it would prove entertaining and that Tupaia would be useful as a navigator and translator. Cook allowed Tupaia and his young son Tayeto on board on the condition that Banks paid for their passage. It was a wise decision, as Tupaia would prove to be an invaluable asset in the months to come.

After visiting the Society Islands, where the party were generally welcomed after an exchange of gifts, the *Endeavour* headed out into the unknown. Conditions on board were bad, with infested food, the onset of scurvy (for which Banks took lemon juice and brandy as a prophylactic) and an outbreak of venereal disease. It must have been with great relief, tinged with trepidation, that after nearly three months land was spotted in the distance on 6 October 1769. As they drew closer, it became apparent that it was a large land mass. Banks thought it could be the fabled south continent, but it was in fact the North Island of New Zealand. During a six-month circumnavigation Cook would prove that the land mass was made up of two distinct islands.

Overleaf: Milford Sound, New Zealand. Banks was impressed by the lushness of the New Zealand landscape, and after his return to Britain often discussed the colonization of the islands.

In contrast to the friendly Tahitians, the Maoris were much more war-like. On one of the first landings to search for fresh water, Banks was startled to hear gunshots coming from the direction of the landing boat. Four natives had attacked the boys tending the boat but had been driven off by the crew. In the skirmish the chief of the war party had been shot dead, and Banks was able to study his first Maori:

> He was a middle sizd man tattowd in the face on one cheek only in spiral lines very regularly formd; he was coverd with a fine cloth of a manufacture totaly new to us … his hair was also tied in a knot on the top of head but no feather stuck in it; his complexion brown but not very dark.

On another occasion, several Maoris were killed when they attempted to steal muskets. Although he believed that the crew had acted in self-defence, Banks deeply regretted the slaughter and felt responsible for it. He wrote in his journal: 'Thus ended the most disagreable day My life has yet seen, black be the mark for it and heaven send that such may never return to embitter future reflection.'

As they sailed on and contacts continued, Tupaia discovered that the Maori language was similar to his own and began making negotiations for provisions and water. One group of Maoris offered a few feathers in exchange for the nails and beads presented to them – probably the first fair exchange of the voyage. As they sailed around the coast Banks began to like and respect the Maoris. He noted the modesty of dress worn by the women, but added wryly that 'they were as great coquetts as any Europeans could be and the young ones as skittish as unbroken fillies'. He was impressed by the sanitary conditions of the villages: each group of houses had a type of lavatory and waste was collected in a dunghill.

There seemed no consistency to the Maoris' response to the *Endeavour* as she made her way around the New Zealand coastline. On several occasions Maori war canoes had to be warned off by cannon fire, and Banks's plant collecting was frustratingly disrupted by the hostilities. Nevertheless, he and Solander were able to make a few forays inland, where they were amazed at the diversity and richness of the flora. All the plants were new to science and Banks was surprised by tree ferns, the world's largest buttercup and a speedwell that grew 39 feet high. Despite the problems in collecting, Banks gathered forty new species including the New Zealand flax (*Phormium tenax*). At Purangi Bay they gathered a large selection of plants, collected wild celery and oysters

Phormium tenax or New Zealand flax. Discovered by Banks, this plant was used by the Maoris to make a type of cloth. It is now widely grown in gardens.

and dined off a number of shags they had shot. The crew found the latter most enjoyable, although Banks admitted that 'hunger is certainly most excellent sauce'. Further up the coast at Hauraki Gulf, he was able to inspect the local trees (*Podocarpus* species) and noted that they were 'the finest timber my eyes ever beheld. Every tree as straight as pine and of immense size.'

Following the North Island coastline, the crew enjoyed a Christmas dinner of freshly shot goose and passed the peak of Mt Egmont on 12 January 1770. A few days later they anchored in Queen Charlotte's Sound, where Banks was able to confirm his suspicions that the Maoris were cannibals. While he was examining a small family camp he discovered two human bones in a provisions basket – they bore the marks of being cooked on a fire and 'the meat was not intirely pickd off from them and on the grisly ends which were gnawd were evident marks of teeth'. The family confirmed that these were the remains of a tribal enemy recently killed in battle. They did not, they stressed, eat their own people.

It was during their stay at Queen Charlotte's Sound that Cook climbed a nearby hill to survey the geography of the land. Seeing the Pacific Ocean, he correctly deduced that there must be a passage through to it (it is now known as Cook Strait), destroying Banks's notion that they were on a large continent. The *Endeavour* then circumnavigated South Island, Banks gamely holding on to his belief until 10 March when they rounded the southern tip and started heading north again.

As the *Endeavour* turned west and headed out into open seas on 1 April 1770, there was time to reflect on a most successful six months. Cook had charted 2,400 miles of coastline and Banks had collected 360 plant specimens. On 17 April 1770 – a historic day – the land of Terra Australis was sighted. At first the new continent (which at one time was going to be named Banksia) appeared disappointingly barren, but as they cruised up the New South Wales coast, fertile hills rose up in the distance. As with New Zealand, the first signs of habitation were columns of smoke rising from the hinterland. On 28 April the *Endeavour* anchored opposite a village, but the natives ignored them. Banks was able to observe the activities of the villagers through his telescope and noted that none of the inhabitants appeared to wear clothing: 'myself to the best of my judgement plainly discernd that the woman did not copy our mother Eve even in the fig leaf.'

A landing party set off in the afternoon but met with fierce vocal opposition from two men. Some muskets were fired as a warning and the men fled into the

This somewhat idealized scene of Cook's arrival in Australia is a wood carving by Samuel Calvert (1828-1913) from a supplement to the *Illustrated Sydney News*, 1865.

bushes, leaving behind their children, hidden beneath a shield and a piece of bark. Banks and his team left the children where they were and threw presents of beads, ribbons and pieces of cloth into the shelters. As a precaution, they removed all the fishing spears from the village before returning to the ship. The following day Banks found the village empty and his presents still lying where he had thrown them. He and Solander spent some fruitful days plant hunting without hindrance and built up a large collection of new species including eucalyptus, acacias, grevilleas, mimosas, the flame tree (*Brachychiton acerifolia*), the crimson-flowered waratah (*Telopea speciosissima*) and the members of the genus named in his honour - *Banksia*. Banks was so impressed by the abundance of plants that he named the area Botany Bay.

The *Endeavour* cruised northwards along the coastline, which varied from barren and desolate to 'woody and pleasant'. With stops at Bustard Bay, Thirsty Sound and Green Island, Banks's plant collection continued to grow at an impressive rate. As Cook sailed through the treacherous coral reefs and islands of the Great Barrier Reef their progress had to be slowed down – often one of the small boats would have to row out in front of the *Endeavour* to sound out the water. Unfortunately, even this prudent form of navigation did not prevent near disaster, for on the night of 10 June 1770 the *Endeavour* struck a reef and held fast. The small boats and the anchor were put overboard as quickly as possible to try to haul the ship off the coral, but without success. They had struck the reef at high tide, and as the water level began to fall the ship became more firmly embedded. At first light the crew could see that they were about 25 miles from the coast and that there were no islands in the vicinity. In order to lighten the *Endeavour,* the water supply, the ballast and the six guns on deck were all thrown overboard. As the tide rose again and the ship started to take in water the three suction pumps were manned. Banks, not given to melodramatics, was clearly convinced that the *Endeavour* was lost and that his chances of survival were not good:

> Now in my opinion I intirely gave up the ship and packing up what I thought I might save prepard myself for the worst. The most critical part of our distress now approached: the ship was almost afloat and every thing ready to get her into deep water but she leakd so fast that with all our pumps we could just keep her free: if (as was probable) she should make more water when hauld off she must sink and we well knew that our boats were not capable of carrying us all ashore, so that some, probably the most of us, must be drownd: a better fate maybe than those would have who should get ashore without arms to defend themselves from the Indians or provide themselves with food, on a countrey where we had not the least reason to hope for subsistance had they even every convenence to take it as netts &c, so barren had we always found it; and had they even met with good usage from the natives and food to support them, debarrd from a hope of ever again seing their native countrey or conversing with any but the most uncivilizd savages perhaps in the world.

To everyone's great relief the ship at last floated off the reef and took in water no faster than before. Even so, the whole crew, Banks included, had to man the pumps constantly for the entire day – an extremely tiring task. The hole in the ship was temporarily plugged with sailcloth filled with wool and oakum, and all thoughts now turned to finding a suitable harbour where repairs could be carried out. Six days later, and with the crew utterly exhausted, the *Endeavour*

An illustration from Banks's *Florilegium*, showing the plant that bears his name –
Banksia integrifolia. The drawing illustrates the skill of the botanical artist.

limped into the mouth of the Endeavour river. When the damaged ship was examined it was discovered that the hole would have 'sunk a ship with twice our pumps' but for a large lump of coral that had broken off and become lodged in the hole.

During the month and a half that it took to repair the ship, Banks explored the countryside and added more specimens to his growing plant collection, including the Moreton Bay pine (*Araucaria cunninghamii*), red cedars, yellow woods, tulip woods, *Hibiscus tiliaceus* and kangaroo grass (*Themada australis*). He also spent time drying out his plant specimens, which had been damaged by bilge water running to the stern of the ship when the *Endeavour* beached – 'many were savd but some intirely lost and spoild'. During this time a kangaroo, giant clams and turtles were caught and eaten. Gradually a relationship was built up with the wary Aborigines, and Banks was at last able to observe them at close quarters. They were, he noted, smaller and darker than the Maoris, went about naked save for a bird's bone through their noses and spoke in a harsh language, which Banks thought was closer to English than any other he had so far encountered. Unfortunately, the early signs of friendship did not last long. When a group of Aborigines was refused one of the turtles on the deck of the ship they responded by returning to the shore and setting light to the surrounding grasses. A quick dash to the beach by Banks saved the few belongings that were there – the ship's supply of gunpowder that had been stored on the shore had fortuitously been taken back aboard a few days earlier.

At the beginning of August the *Endeavour* was finally shipshape again, but it was some nine days before a suitable passage through the treacherous reef to the open sea – Cook's Passage – was found. Finally, on 14 August the ship was out of sight of land for the first time in nearly four nerve-racking months, and Banks could look with pride at the 331 Australian plants he had gathered.

The *Endeavour* had now been at sea for two years, and crew members who had wanted to set off for home after the New Zealand leg were now even more homesick and restless. Cook was also keen to reach Batavia (now Jakarta) on the Indonesian island of Java, and after another close scrape with the Great Barrier Reef and a display of aurora australis (an unusual event so far north), they arrived there on 9 October. The crew's delight at being ashore in 'civilization' was quickly dampened – fever was rife in the unhealthy town, with malaria-carrying mosquitoes breeding prolifically in the stagnant canals, and soon many of the crew, including Cook, Banks and Solander, were suffering from the initial symptoms. Monkhouse, the ship's surgeon, was one of the first to die, followed by Tupaia and little Tayeto. The initial relief at leaving the town on 26 December was tempered by the frequent recurrence of fever among the crew. From 23 January 1771 Banks recorded the death of one crew member every day for the next seven days. By the end of the voyage the *Endeavour* had lost forty-two of her ninety-four crew members, and of Banks's party only he and two others had survived. The vast majority of casualties were victims of the two-and-a-half-month stay in Java.

A little over two months later, the *Endeavour* arrived off the coast of South Africa. Although he noted the beauty of the heaths (heathers), Banks does not appear to have spent any time plant collecting during the month the party stayed at Cape Town. The ship left South Africa on 14 April and after a brief stop off at St Helena, where Banks recorded his disgust at the ill-treatment of the slaves, they sailed into British waters on 10 July 1771. When we think of the *Endeavour* voyage today it is Captain Cook's name that is remembered, not that of Banks. However, in 1771 it was Banks who arrived back to a hero's welcome and was lauded in society circles. Cook's contribution to the success of the voyage was generally overlooked. (He went on to command two subsequent circumnavigations, but on the second of these he was hacked to death by natives in Hawaii.)

Banks now set about making a detailed study of his huge collection. When all the dried herbarium specimens had been described, catalogued and named, they amounted to a staggering 1,300 new species and 110 new genera. Although some of these later became garden favourites, for example *Gaultheria mucronata*, the everlasting flower (*Helichrysum bracteatum*), *Grevillea glauca* and *Hebe elliptica*, his discoveries did not have a significant impact on the late eighteenth-century garden. This was partly because he did not bring back viable seed or live plants, but also because at heart he was a scientist, seeing plant hunting only in terms of pushing forward the understanding of the natural world. Nevertheless, he did play an indirect role in the development of garden fashions through his establishment of Kew and his policy of organized plant hunting.

Banks had grand plans for botany and science in general, and needed a substantial but centrally located house in London to serve as his headquarters. In March 1777 he purchased 32 Soho Square, where he lived with his unmarried sister Sophia and his wife Dorothea Hugessen, whom he married on 23 March 1779. The house in Soho Square housed Banks's enormous herbarium and extensive library, and throughout the chaos of war in Europe it remained a

Joseph Banks, aged thirty, painted by Joshua Reynolds (1772–3).

haven where scientists of all nationalities could meet freely and informally, discuss and debate, and enjoy one of Banks's famous working breakfasts. His conviction that science should be apolitical and his sense of fair play were so strong that on several occasions when French ships were captured and found to contain scientific collections he insisted on their repatriation rather than keeping the spoils for himself.

To provide a retreat from the city, Banks leased Spring Grove, Heston (3 miles from Kew and 10 miles from Soho Square), from Elisha Biscoe in 1779 and finally bought it in 1808. He extensively modified the grounds, erecting greenhouses, a grape house, a peach house and an ice house, and making large soft fruit and kitchen gardens. He turned the 49 acres into an experimental station for plant (especially fruit) and animal breeding. (He had strong interests in sheep-breeding and smuggled merino sheep from Spain, later introducing them into Australia.)

Banks did not simply pursue science for his own ends. As well as being called a 'father of Australia' for his promotion of its colonization, he also worked hard to establish Kew as the world's premier botanic garden. Today keen gardeners everywhere associate the name of Kew with the beautiful landscape and glasshouses, a tranquil oasis of plants, only a stone's throw from the centre of London. Its history goes back to the 1660s, when Sir Henry Capel collected rare and exotic plants at Kew House. In 1730 Frederick, Prince of Wales, obtained the lease and employed the famous landscape designer William Kent to make him a new garden. When Frederick died in 1751, his widow, Augusta, the Dowager Princess of Wales, herself a keen gardener, teamed up with the Earl of Bute and with his help added the grounds of neighbouring Richmond Lodge (sometimes called Richmond Palace). The pair were also responsible for inaugurating the scientific study of botany at Kew when in 1759 they engaged William Aiton to lay out a physic garden covering 8.9 acres. When Augusta's son, George III, became king the following year at the tender age of twenty-two, he encouraged his mother's passion for gardening. During the next two years he built her one of the biggest stove-houses (hot-houses) in the country and added a large collection of trees. After her death in 1772, the King bought Kew House, and it remained one of his favourite royal residences and a refuge during his periodic bouts of insanity (he suffered from porphyria).

Banks's *Endeavour* exploits came to the King's attention and he invited him to Kew in 1771. Sharing a passion for plants, Banks quickly gained royal favour. In 1772 the Earl of Bute, whom the King disliked, was banished and in his place Banks was appointed Scientific Adviser on the Plant Life of the Dependencies of the Crown – a sort of unofficial director of Kew. The next year the King employed the most fashionable garden designer of the time, Lancelot 'Capability' Brown, to lay out the new garden, and the King and Banks were often seen strolling together there. Banks wanted to organize a scientific study of the flora of all the British colonies, but he needed two things: a location to grow, study and store these plants, and plant material for study. To make his vision become reality, Banks used his position of favour with the King. He persuaded George that he should have the world's most diverse collection of plants and that the logical location for this collection was the Royal Garden at Kew.

Under Banks's careful manipulation Kew was slowly transformed from a royal pleasure ground into a research-oriented botanic garden. He used his broad network of social contacts, his reputation within the scientific community and his friends to ensure a supply of new plants from around the world, backing up this somewhat random strategy by sending educated botanists to remote, unexplored parts of the Crown colonies with specific instructions to find and bring back new material. This latter policy is the reason so many wonderful garden plants arrived in Britain and why so much controversy has arisen about Banks. For as well as ornamental introductions, the plant hunters sent back economic crops, and Banks quickly saw the financial benefits of exchanging commercial crops among colonies. His single-handed establishment of economic plant transfers played a vital role in Britain's emergence as a world power. It also led directly to the exploitation of human and natural resources in the colonies.

It was not all plain sailing: one of Banks's schemes involved transporting breadfruit plants from Tahiti to the West Indies in order to provide a cheap food for the slaves. Unfortunately the plant hunter David Nelson, charged with the transfer, set sail upon a certain HMS *Bounty* under the command of Captain William Bligh.

With his hectic town life Banks needed times of peace and tranquility, and during the autumn hunting season and later in life when he suffered from periodic attacks of gout he moved his household back to Revesby. He was a frustrated chair-bound invalid from 1810 but persevered with his work and vital correspondence and did not relinquish the Presidency of the Royal Society (which he had held for forty-two years) until 1 June 1820, eighteen days before he died.

Through his travels, studies and patronage of scientists Banks proved that there was more to him than wealth and powerful connections. His social status, including the friendship of a king, was used wisely and for the greater good of science rather than for personal glory. He established Kew as a world-renowned centre of excellence, a position it retains to this day, and although his plant transfers brought misery to others, this was not his intention. Perhaps his greatest gift to the gardener was the creation of a systematic and world-wide plant hunting programme that ensured that throughout the ensuing two centuries thousands of wonderful new plants would become available across the world. Through his work at Kew, Banks is credited with securing the introduction of over 7,000 new species, and in the next chapters we shall witness the extraordinary results of this policy to send out professional plant hunters.

Joseph Banks's Plant Introductions

The date given beside each plant name is the date of its introduction into Britain.

Leptospermum scoparium (1771)

Leptospermum = (Gk) *leptos*, slender; *sperma*, seed
scoparium = (Lat.) broom-like

Long-lasting, white to pink, button-like flowers crowd the upright stems in early to mid-summer. The manuka or tea-tree is a variable evergreen large shrub or small tree, with small, purple-green aromatic leaves. Many colour forms exist, e.g. 'Red Damask' (1944, deep red), 'Nicholsii' (*c.* 1926, carmine red with bronze-purple foliage) and 'Keatleyi' (soft pink).

From New Zealand, Australia and Tasmania, it is common in many habitats, reaching 26ft (8m) by the coasts or becoming prostrate in the mountains.

Sophora tetraptera (1771)

Sophora = from the Arabic name
tetraptera = (Gk) *tetra*, four; *ptera*, wing, after the
 seed-pods

Elegant, clear yellow, tubular pea-shaped flowers hang in drooping clusters from July to September. The kowhai is a large shrub or small tree, the spreading branches bearing attractive 8–15in (20–40cm) pinnate green leaves and four-winged seed-pods in autumn. The very similar species *S. microphylla*, also found by Banks, occurs on both North and South Islands and has smaller flowers and leaves. The form *S. m.* 'Sun King' is very hardy.

New Zealand's national flower. It occurs in forest fringes, damp woods and rocky areas of the North Island, reaching 26ft (8m) tall.

Banksia integrifolia (1788)

Banksia = after Sir Joseph Banks
integrifolia = (Lat.) entire leaves

Wonderful cone-like spikes crowded with yellow flowers adorn dense bushes in autumn. The Australian honeysuckle is an evergreen shrub with handsome leathery olive-green leaves, white beneath. The genus was named after Banks and contains about seventy species of extraordinary tender shrubs and trees, all from Australia. Other species include *B. serrata*, with silvery-grey flowers, *B. occidentalis* (swamp banksia), with orange flowers, and *B. speciosa*, with yellow-green flowers.

From eastern Australia, it reaches 6½–10ft (2–3m) in height.

Callistemon citrinus (1788)

Callistemon = from (Gk) *kalli*, beautiful;
 stemon, stamen
citrinus = (Lat.) lemon-coloured

Exotic, brush-like spikes of vivid crimson tufted flowers top arching stems in summer. The bottle-brush is a vigorous spreading evergreen shrub with narrow, rigid foliage, which is lemon-scented when crushed. A hardier form is *C. c.* 'Splendens', with brilliant scarlet flowers all summer, reaching 5–6½ft (1.5–2m). Banks also introduced *C. linearis* (1788), with long scarlet brushes, and *C. salignus* (1788), a hardier species with pale yellow flowers.

From eastern Australia, growing in damp areas and reaching 3–10ft (1–3m) in height.

Phormium tenax (c. 1789)

Phormium = (Gk) mat (from its fibre)
tenax = (Lat.) strong, tough

Imposing clumps of sword-like, leathery grey-green leaves, up to 10ft (3m) long. The New Zealand flax is a bold architectural evergreen, which produces unusual 13ft (4m) tall flower spikes bearing bronze-red blooms in July and August. Coloured leaf cultivars include 'Sundowner' (striped cream and pink, purple leaves, less hardy), 'Veitchii' (striped cream and yellow, green leaves) and 'Purpureum' (6½ft/2m; long, bronzy-purple leaves).

From New Zealand and Norfolk Island (naturalized in west Eire, the Azores and elsewhere), growing to 10ft (3m) tall in lowland marshes.

Callistemon citrinus, one of the bottlebrush family, native to Australia. Their architectural flowers provide a flash of crimson in summer.

2

TO THE FAIREST CAPE

Francis Masson

(1741–1805)

WHILE JOSEPH BANKS WAS MEETING WITH GEORGE III AND PERSUADING him to build up a collection of plants at Kew, Captain Cook was preparing for his second circumnavigation. Initially, Banks had planned to accompany Cook, but after a disagreement with the Admiralty (part of which centred on the Admiralty's refusal to allow Banks to take with him a pack of greyhounds and his personal orchestra), Banks arranged for his place to be taken by Kew's first official plant hunter, Francis Masson.

After three and a half months at sea, Cook piloted HMS *Resolution* into Table Bay at Cape Town, South Africa. It was 30 October 1772, and for

---

Left: Morning in the Karoo. The harsh landscape made travelling difficult, but the fascinating plants were a worthy compensation.

Above: Francis Masson, the first official plant hunter sent out by Joseph Banks from Kew. In 1772 he visited South Africa, and he spent the next thirty-four years collecting plants on three continents.

Masson, uprooted from his safe, quiet job as an 'under-gardener' at Kew and transplanted to this emerging colony (at that time under the control of the Dutch East India Company), spring was about to blossom for the second time that year. The reversal of seasons in the two hemispheres was just the first of many adaptations the unworldly thirty-one-year-old from Aberdeen would have to make in the taxing years to come. If he was daunted by South Africa, with its subtropical climate, wild and rugged scenery, exquisite plants and ferocious animals, he was to prove himself up to all the hardships and diffi-culties that came his way.

Masson spent the first few weeks becoming acclimatized and befriending the local inhabitants, the indigenous Hottentots and the Dutch settlers, but it was not long before he experienced the danger that lurked just inland. On one of his first excursions to explore the looming mass of Table Mountain, Masson became entranced by his plant collecting, losing both sense of direction and track of time. He had been warned that a gang of escaped convicts was on the run in the area, and as dusk began to fall the sound of male voices and clanking chains disturbed the quiet. Masson was jolted out of his reverie and, knowing that to be caught by the desperate runaways would most likely mean death, he spent a terrifying evening hiding in the shrubby undergrowth. Eventually he found shelter in an old shepherd's hut, only to discover that the door would not shut properly. He had only a clasp-knife with which to defend himself, and knew he was still in grave danger. After a nerve-racking night spent curled up on the floor, dawn finally broke and he was able to slip away as the first light softly lit the mountainside.

It appears that Masson was made of sturdy stuff, for he was not in the least bit taken aback by this scare and was eager to get on with his mission proper – to explore the interior. On 10 December 1772 he set off eastward in a hired covered wagon pulled by eight oxen and accompanied by its native driver and a Scandinavian mercenary called Franz Per Oldenburg, who acted as guide and interpreter. Although this was a fairly short round trip of 400 miles across the Cape Flats, taking in Paarl, Stellenbosch, the Hottentots Holland Mountains, the hot springs at Swartberg and Swellendam, it provided an excellent foretaste of future expeditions. Masson experienced for the first time the difficulties associated with travelling by cart, such as fording treacherous rivers and tra-versing rough paths, but he was rewarded with a wonderful display of the veld flora at its most beautiful. It was on this trip that, as he later wrote: 'I collected seed of so many beautiful species of erica which have succeeded so well in the

Royal Garden at Kew.' On the Hottentots Holland Mountains Masson found the Cape heaths, and recorded in his journal for 5 January 1773: 'These mountains abound with a great number of curious plants, and are, I believe, the richest mountains in Africa for a botanist.' He finished the same day's entry on a more practical note, observing that he 'lodged at a mean cottage and the Dutch and the Hottentot lived almost promiscuously together, their beds consisting only of sheep skins'.

Masson arrived back in Cape Town at the end of January, knowing Banks's assertion that the province would provide rich botanical pickings was correct. For the next few months he spent time learning more about the nature of the countryside from the Dutch farmers and planning a grander expedition. News of this proposed expedition came to the ears of the Swedish botanist Carl Per Thunberg, a pupil of the Linnean School at Uppsala, Sweden, and he persuaded Masson that they should undertake the trip together. It was a peculiar friendship. The two men's characters were completely opposite – Thunberg was an incorrigible braggart and show-off, while Masson was quietly spoken and modest. Nonetheless, working as a team they proved to be exceptionally efficient and successful, as can be seen by the fruits of Masson's second excursion into the hinterland.

They set off on the north-west coast route to the Blaauwberg (Blue Mountain) on 11 September 1773, with an ox-wagon full of supplies and collecting equipment, a European servant and three Hottentot drivers-cum-helpers. Masson and Thunberg had sensibly opted to travel on horseback rather than walk, which not only saved their energy but also meant that they could make excursions away from the supply vehicle. During the first week the weather was damp and overcast, but the botanists paid little attention as both were busy studying the summer flora. Masson recounts that he was 'delighted to see the luxuriance of the meadows, the grass reaching to our horses bellies, enriched with a great variety of ixiae, gladioli and irises, most of which were in flower at the Cape in the month of August'. The next day they reached Saldanha Bay, where Masson found *Amaryllis disticha,* 'which the Dutch call vergift-boll, poison bulb; the juice of which they say, the Hottentots use as an ingredient to poison their arrows'. Masson was continually surprised by the variety and splendour of the flowers, writing in his journal on 27 September: 'The whole country affords a fine field for botany, being enamelled with the greatest number of flowers I ever saw, of exquisite beauty and fragrance.' The following day he added the pretty green-flowered bulb *Ixia viridis* to his tally.

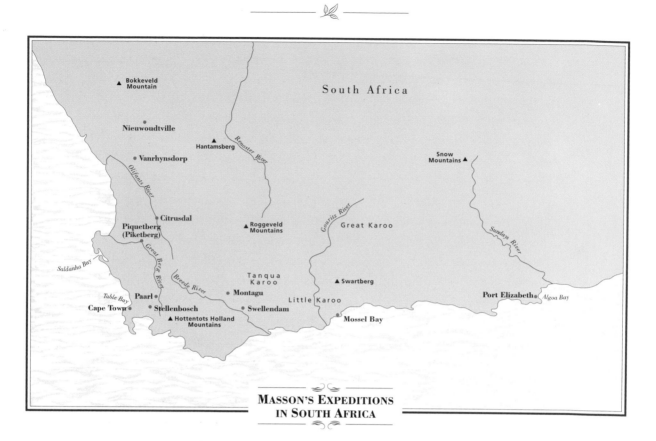

MASSON'S EXPEDITIONS
IN SOUTH AFRICA

Turning eastward, the team successfully forded the flooded Berg river and headed towards the Karoo plateau beyond it. The precarious ascent 'by a passage called Kartouw, the most difficult in the Cape Province', with steep drops on either side, was made still more dangerous by poor weather. For three long, anxious, rain-soaked hours Masson and Thunberg had to gently lead their terrified and stumbling horses while the wagon swayed dangerously on the edge of the precipices behind them. Exhausted, they eventually made it through the pass and descended in a thoroughly bedraggled state to a Dutch settler's hut. Although the settler was hospitable, it was not the most comfortable of places to spend the night – there was just one room split into compartments by reed mats – but, as Masson pragmatically observed, 'cold and wet as we were, we were glad of anything'.

On 10 October the party crossed the Olifants river, following its course to Citrusdal. Here, 'after some warm debates', Masson and Thunberg left the wagon to be repaired and set off on horseback into the barren and extremely rugged mountain range to the north-east. After the discomfort of heavy rain, the two botanists were now subjected to the searing South African sun beating relentlessly down on them, but their perseverance was rewarded, for among the desiccated landscape they discovered *Protea grandiflora* and a number of splendid succulents such as stapelias and massonias. Once reunited with the supply wagon, the team followed the grassy banks of the Breede river before passing through the rolling hills near Montagu and arriving on the plains at Swellendam. Although they had travelled some 300 miles across the tortuous land, they were only 200 miles from the Cape as the crow flies.

After a much needed rest, Masson and Thunberg headed east with the intention of exploring the Little Karoo. On 10 November, however, disaster almost struck when they were attempting to cross the swollen Duvvenhoek's river. The differing accounts of what happened next provide an interesting contrast in the men's characters. Thunberg records that he, being 'the most courageous of any of the company, and, in the whole course of the journey, was constantly obliged to go on before and head them, now also, without a moment's consideration, rode plump into the river'. His horse fell into a deep hippopotamus pit on the river bed and floundered for some minutes before gaining the opposite bank. In his account, Masson mentions none of the astonishment and dread that Thunberg alleges he felt while watching, but tersely wrote, 'the Doctor imprudently took the ford without the least enquiry,' and implied that it was the horse's strength that saved the puffed-up Swede rather than any staggering feat of horsemanship. Hippopotamus pits were a frequent hazard encountered during river crossings, although there were only a few of the animals left within 800 miles of Cape Town. The Boer farmers had killed so many for their pork-like flesh and their hides that a shooting ban had been imposed. The party forged on unperturbed by this misadventure, crossing the Gouritz river, which rose up to their saddles, before reaching the sea at Mossel Bay on 16 November. Turning north, they now crossed the Attaquas Pass (2,000ft) and entered the very inhospitable land of the Little Karoo, an area which Masson described thus:

> no land could exhibit a more wasteful prospect, the plains being nothing but rotten rock intermixed with a little red loam in the interstices supporting shrubby bushes, evergreens but by the scorching heat of the sun, stripped almost of all their leaves. But we found new succulents never seen before which appeared like a new creation.

Beyond this wasteland, Masson and Thunberg spent more than a month exploring the ever-changing countryside, with its rich botanic variety. Crossing verdant plains, they passed through hilly woodland and traversed narrow valleys. Occasionally they encountered a remote settlement, and were somewhat surprised to discover that some of the farms were quite rich. Masson, however, was dismayed by the treatment of the blacks by the Dutch farmers, 'who give them for wages beads, and tobacco mixed with hemp; the latter, which intoxicates them, they are extremely fond of'.

By 14 December they had reached Algoa Bay, near Port Elizabeth, some 489 miles from the Cape, and on the 17th they reached Sundays river where Masson's local guides refused to continue into the Snow Mountains. This was not for any sacred reason, but simply that the mountains were the territory of a particularly fierce tribe of Hottentots who would, they warned, kill them all just for the iron on the wagon. It was on this leg of the journey, as they progressed into more remote areas, that the party suffered its first serious brush with the local wildlife. A pack of hyenas attacked the wagon, injuring one of the oxen, and was only just beaten off. Lions and buffaloes became more numerous, and evasive action often had to be taken.

By now the oxen were in a poor state of health, and Masson reluctantly turned back towards Cape Town, travelling parallel with the coastline. This route took them over the high ridge of the Lange Kloor, where on 30 December Masson found *Erica tomentosa* on a mountain ledge. He also found *Stapelia euphorbioides* and *Geranium spinosum* near the Great Thorny River. Their adventures were not quite over yet, for twice on the return journey Masson and Thunberg left the ox-wagon to go exploring and became thoroughly lost. On both occasions they were forced to spend an extremely uncomfortable night fighting off the cold – with only sufficient fuel to keep a meagre fire burning, they were forced to pace up and down all night to keep warm. After four and a half exhausting months, covering 1,000 miles, the party arrived back at Cape Town on 29 January 1774. Among the new plant species secure in the ox-wagon was the striking bird-of-paradise flower (*Strelitzia reginae*). The expedition was a botanical triumph, and Masson busied himself over the next few months with sorting and dispatching his collection back to Kew. Perhaps as a break from the work, Masson and Thunberg went on a few short plant hunting trips in the local vicinity and spent some time escorting Lady Ann Monson, an aristocratic botanist of considerable note, around the Cape.

Full of confidence, and with the southern hemisphere spring in the air, Masson began to make preparations for another expedition into unknown territories. On 25 September 1774 he set off, with two Hottentot servants to drive his ox-wagon and tend to his horse. After several damp days crossing the coastal plain, Masson caught up with the irrepressible Thunberg at Paarl. They spent some time collecting among the local peaks before heading north across scrubland studded with glorious spring flowers. Crossing the rain-swelled Berg river with difficulty, on 13 October the expedition reached the foot of the Piketberg (then Piquetberg), where among many botanical gems they found *Stapelia incarnata*.

As they moved north again, in the direction of the source of the Olifants river, they entered an arid, desert-like landscape. The boiling sun made travelling possible only in the early morning and evening, and even then progress was slow. The sandy soil was pitted with mole-rat burrows, causing the horses to stumble dangerously every few minutes, and the countryside was alive with poisonous snakes, which, Masson recorded, slithered among the hooves of the pack animals and coiled themselves around the men when they stopped to rest. The threat of heat exhaustion and lack of water were continual problems, and at one point the oxen became so dehydrated that Thunberg feared they would keel over and die. Despite the apparent barrenness of their surroundings, however, Masson managed to find a number of succulents including mesembryanthemums, euphorbias and stapelias. To their great relief, the group reached another isolated Dutch farmstead on 25 October and were able to enjoy a short period of relaxation.

At the end of October the party crossed the Olifants river by boat and a few days later came to Vanrhynsdorp, where they began the three-day trek to the Bokkeveld Mountains. They were overtaken by

The striking bird-of-paradise flower, *Strelitzia reginae*. Perhaps the loveliest of Masson's introductions, it remains a popular conservatory plant and cut flower.

a Boer farmer, who warned them that there was no water source visible from the track itself but said that he would tie a rag to a tree to indicate where water could be found a little way off. As they trudged ever further into the boulder-strewn desert it became increasingly obvious that the farmer had forgotten his promise. Again the poor oxen gradually grew weaker and weaker beneath the merciless sun, and Masson sadly records that he could collect only the succulents that grew beside the track. Even so, he managed to find 100 species, including the stone-like *Lithops* species. Relief came when the Boer farmer, having reached his home, sent two fresh sets of oxen back to help the struggling Europeans over the mountains surrounding the Karoo. This kind gesture was invaluable because the 2,000-foot ascent of Bokkeveld Mountain was made on an extremely rugged and steep path, and five Hottentots straining on ropes were needed to prevent the wagon from overturning. On the summit they were

Protea cynaroides, the national flower of South Africa. Its bizarrely shaped flowers appear in early summer.

revived by the cool mountain breezes and celebrated the discovery of the 13-foot tall *Aloe dichotoma*. The party continued northwards along the plateau before descending to the dry, dusty plain near Nieuwoudtville and travelling north-east over 'rotten rock' to Hantamsberg. Lying some 350 miles north of Cape Town, this was then the most remote Dutch settlement in existence.

Turning south-east, Masson now crossed further tracts of dry land before reaching and following the Renoster river, arriving at the Roggeveld Mountains in mid-November. On the 16th the party once again had to ascend a treacherously difficult path before reaching the veld at 4,000 feet above sea-level. The veld stretched for several hundred miles along the ridge of the mountains, and in stark contrast to the incessant heat of the Karoo, the party was now buffeted by icy storms. After spending two days under cover waiting for one particularly ferocious gale to blow over, Masson and Thunberg decided they had had enough. They abandoned their plans to travel north-east and descended to the plain below. This change in plan presented its own dangers, however, as Masson described in his journal:

> We were furnished with fresh oxen, and several Hottentots, who, with long thongs of leather affixed to the upper part of our waggons, kept them from overturning, while we were obliged to make both the hind wheels fast with an iron chain to retard their motion. After two and a half hours employed in hard labour, sometimes pulling on one side, sometimes on the other, and sometimes all obliged to hang on with all our strength behind the waggon, to keep it from running over the oxen, we arrived at the foot of the mountain, where we found the heat more troublesome than the cold had been at the top.

Exchanging frost for furnace, Masson and Thunberg now endured four thirsty days crossing the Tanqua Karoo, frequently travelling at night in an attempt to alleviate some of the discomfort. On arriving at the Bokkeveld Mountains in mid-December, they were delighted to find a stream of cool water where they 'spent the night and part of the next day in luxury'. Thoughts of the comforts of Cape Town must have begun to lighten the weary travellers' minds, and the little party entered Cape Town on 29 December – just in time for some well-deserved Hogmanay celebrations.

Masson, however, had no time for an extended holiday. He spent the following months writing-up his journal and botanical observations and dispatching his latest collection of new species. This time there were over 500, including *Amaryllis belladonna* and *Protea cynaroides*. Banks was ecstatic, for he could now go to the King with unequivocal evidence that Kew was developing the world's best botanical collection, and would be able to send out more plant hunters. Meanwhile, Masson was recalled and arrived back in Britain at the end of 1775. The success of his collecting can in part be judged by a letter written on 5 May 1776 by the Reverend M. Tyson: 'Mr Masson showed me the New World in his amazing Cape hothouse, erica 140 species, many proteas, geraniums and cliffortias more than 50.' In the same year Masson wrote self-deprecatingly to Linnaeus declaring that he had found over 400 new species, and in an article in the *Philosophical Transactions of the Royal Society* he was the first to use the now-familiar title 'The Royal Botanic Gardens at Kew'. Not a man happy to bask in attention or to live off past glories, he found it hard to re-adjust into the routine of Kew after such an exhilarating time in South Africa. He continually begged Banks to find him another expedition, and on 9 May 1778 he set off on a transatlantic circuit, beginning in Madeira, Tenerife and the Azores before moving on to the West Indies.

Given the ill luck that Masson was to encounter, it is not surprising that no journal exists for this expedition; it is clear though, that on the early leg of his journey he successfully collected many plants. The first dispatch of sixty plants from Madeira reached Banks in July 1778, and in May 1779 he received a further 123 species. Masson's finds from the Canary Islands included the spectacular echiums, with their rocket-like spikes of blue flowers that reach 10 feet plus, and *Senecio cineraria*, which would become the parent of today's colourful and popular cinerarias. Unfortunately, trouble awaited Masson in the Caribbean. On 2 July 1779 French troops invaded Grenada, and he was caught in the wrong place at the wrong time. He was forcibly conscripted into the local militia and made to defend the main town and harbour. He was captured and imprisoned by the French, lost his plant collection and was released only after high-level negotiations by Banks. Bad luck continued when he moved on to St Lucia: a hurricane devastated much of the island in October 1780 and Masson's new plant collection, equipment and journal all ended up among the wreckage. Nothing could be saved, and in a state of utter dejection he sailed for home, arriving back in early 1781.

In 1783 Masson set off for a two-year exploration of Lisbon (where he tried his hand at garden design), Portugal and Algeria. After a brief return to Kew, he

Right: A letter dated 26 December 1775, written by Masson to the great Linnaeus, in which he states that his two South African adventures have yielded 'upward of 400 new species'.

Honourable Sir

Printed in Linn. Corresp.
v. 2. 559.

I Hope your goodness will excuse the Liberty
I have taken in addressing myself to you, as it proceeds from
a knowledge of your superior Merit, and your exalted character in Natural History. I have been employed some years
past, by the King of great Brittain in collecting of Plants
for the Royal Gardens at Kew, my researches have been
chiefly at the Cape of good Hope, where I had the fortune
to meet with the ingenious Docter Thunberg; with whom I
made two successfull journies into the interior parts of
the country; My labours have been crowned with success,
having added upwards of 400 new species plants to his Majesties collection of living plants, and I believe many new Genera.

I expect soon to go out on another expedition, to another part of the glob, to collect plants, for his Majesty,
and if I should be so fortunat to discover any thing New in
any branch of Natural history I should be happy in having
the honour of communicating it to you. I had the pleasure
of seeing Mr Sparrmann at the Cape and received from him
a parcel of Seed which he collected in the Southeren Islands
which I now send you, I would not presume to send you any cape
Plants as I presume Dr Thunberg has sent you every
kind that he hath collected which are much the Same with
mine.

left London aboard the East Indiaman *Earl of Talbot* in late 1785 and headed for his favourite plant hunting location. He arrived at Cape Town on 10 January 1786 to find the colony a place of suspicion and mistrust. As Britain and Holland were now at war, the Dutch authorities had decreed that foreign visitors were not allowed to travel any further than a three-hour walk from Cape Town. British citizens were under particular scrutiny, as they were considered to be potential spies. Masson's passion for discovering plants, his sense of loyalty to his patron and his love of excitement overrode any misgivings he may have had in blatantly ignoring the Dutch edict. In March he sent Banks seed of 176 species, including the arum lily (*Zantedeschia aethiopica*) and more Cape heaths. During the next eight years the now middle-aged Masson made a number of long treks into the interior (Banks actually admonished him over his wanderlust and urged him to concentrate on particular areas). He collected more botanical delights for Kew, which he carefully cultivated in a small garden in Cape Town before shipping them home. Finally, the pressures of the political unrest became too much to bear, and he returned to England in March 1795.

Having spent over twenty years of his adult life exploring exotic and dangerous terrain, Masson was now completely unable to settle down to a life of tranquillity among the greenhouses of Kew. The call of the wild beckoned once more, and he set sail for North America in September 1797. The crossing was unpleasant. The ship was boarded twice by French pirates and on the second occasion the passengers were forced onto a Bremen vessel bound for Baltimore. Conditions on board were atrocious: they had to subsist on half a pound of unappetizing bread a day, drink dirty water and sleep among the ropes. To add insult to injury, the weather was foul throughout the latter part of the journey. After a reviving stay at New York, Masson headed north, eventually crossing the border into Canada. He sent consignments of seed back to Kew, including the lovely wake robin (*Trillium grandiflorum*), but he was no longer a young man and found the harsh climate around Montreal debilitating. Used to the oppressive heat of South Africa, Masson's constitution deteriorated during the bitter winter of 1805 and he died on 23 December, thousands of miles from his home and friends.

Right: The elegant flowers of the white arum lily (*Zantedeschia aethiopica*) were fashionable in Victorian gardens and are still popular today.

Gardeners owe a large debt to Masson for his introduction of many brightly coloured flowering plants that are still enjoyed today. Gladioli, amaryllis, cinerarias, streptocarpus, the bird-of-paradise and proteas lighten our rooms and conservatories either as indoor plants or cut flowers. The garden flower border would be poorer were it not for Masson's kniphofias (nine species), agapanthus, and the flamboyant *Zantedeschia aethiopica*. Bedding displays and hanging baskets are brightened by his pelargoniums (forty-seven species) and lobelias. South Africa's colourful summer- and autumn-flowering bulbs provided rich pickings for Masson, and we still enjoy the sparaxis, ixias, watsonias, haemanthus, tulbaghias and oxalis that he introduced.

Although he did not bring back all the Cape species which were to prove horticulturally important, Masson's pioneering work spawned the great interest in region's flora which stimulated others to follow in his footsteps and which continues to this day. Perhaps his most remarkable and endurable introduction is the cycad, *Encephalartos alternsteinii*. His original introduction from 1775 is still thriving in the Palm House at Kew, making it probably the oldest pot-plant in the world!

Francis Masson's Plant Introductions

The date given beside each plant name is the date of its introduction into Britain.

Strelitzia reginae (1773)
Strelitzia = after Charlotte Sophia of Mecklenberg-Strelitz, later Queen of George III
reginae = (Lat.) of the Queen

Exotic bright orange and indigo-blue flowers form an upright, pointed crest above boat-shaped green and red bracts on 4ft (1.2m) stems in April and May. The aptly named bird-of-paradise flower is perhaps the quintessential tropical bloom. It produces a fan-like rosette of evergreen 18 in (45cm) long greyish-green, leathery, paddle-shaped leaves.
From South Africa, growing in coastal woods in the Eastern Cape and reaching 5ft (1.5m).

Protea cynaroides (1774)
Protea = after the Greek sea-god Proteus, who assumed diverse shapes
cynaroides = resembling (Gk) *Cynara*, the artichoke

Spectacular 8–12in (20–30cm) wide artichoke-like flowers, with rows of flesh pink, overlapping stiff bracts surrounding a dome of silky-haired pinkish-white stamens in May and June. The king protea or honeypot sugarbush, a remarkable evergreen shrub with 5–6in (12–15cm) long thick, wavy-edged leathery leaves and red stems, is the most striking species.
From South Africa, growing in various habitats from coastal scrub to mountain slopes at 3,200ft (1,500m) in the south-west, south and eastern Cape region, reaching up to 6½ft (2m) tall.

Amaryllis belladonna (c. 1774)
Amaryllis = after the beautiful shepherdess of classical and English poetry
belladonna = (Ital.) beautiful lady; the juice was used to give brilliancy to the eye (it dilates the pupil)

Wonderful deep pink, funnel-shaped, fragrant flowers on thick 20–30in (50–80cm) stems in autumn. The belladonna lily is a glorious bulbous perennial, producing clumps of 12–18in (30–45cm) long straplike leaves after flowering.
From South Africa, growing in scrub, rocky slopes and near rivers in the Cape Province.

Zantedeschia aethiopica (c. 1796)
Zantedeschia = after Francesco Zantedeschi, an Italian botanist
aethiopica = (Lat.) African

Purest white, waxy, 5–8in (12–20cm) funnel-shaped spathes wrap elegantly around golden yellow spadices (flower spikes) on 4½ft (1.5m) stems from March to June. The white arum lily is a beautiful evergreen perennial with 12in (30cm) arrow-shaped dark green leaves emerging from a tuberous rhizome. The variety 'Crowborough' is hardier.
From South Africa, growing in marshy places from the Cape to the Drakensberg Mountains.

Above: The belladonna lily (*Amaryllis belladonna*) is one of the joys of the autumn garden.

3

A WALK ON THE WILD SIDE

David Douglas

(1799–1834)

THE TREATY OF VERSAILLES, SIGNED IN 1783, BROUGHT TO AN END the American War of Independence and signalled the birth of a new nation. In the late eighteenth and early nineteenth centuries, however, 'America' extended west from the eastern seaboard only as far as the Mississippi river and as far north as the Great Lakes. Vast tracts remained almost unexplored and were claimed variously by the British, American and Spanish authorities. The new English-speaking nation quickly began to forge

Left: The grand fir (*Abies grandis*) helped change the British landscape in the nineteenth century. It is shown here in its juvenile form, which is how many Victorians would have seen it, rather than the 200-foot giants Douglas encountered in the wild.

Above: David Douglas, in a crayon drawing which is attributed to Sir Daniel McNee, looking somewhat more dapper than when he was collecting plants in north-west America.

political and trade links with Europe, and as pioneers began to push further and further west, new plants found their way back to Britain.

These were not the first plants from the New World. A steady trickle had been arriving ever since John Tradescant the elder secured several introductions through his subscription to the Virginia Company, namely Virginia creeper (*Parthenocissus quinquifolia*), *Aquilegia canadensis* and *Tradescantia virginiana*. His son visited eastern North America three times during the seventeenth century and is credited with introducing the swamp cypress (*Taxodium distichum*, the tulip tree (*Liriodendron tulipifera*), the butterwood (*Platanus occidentalis*), *Yucca filamentosa* and *Anaphalis margaritacea*. John Banister (1654–92) introduced the first magnolia (*Magnolia virginiana*) in 1688, John Bartram (1699–1777) discovered the first large-leaved rhododendron (*R. maximum*), and in the late eighteenth century John Fraser (1750–1811) discovered *Rhododendron catawbiense* on Great Roa Mountain.

It was Archibald Menzies (1754–1842) who first drew attention to the potential botanical bonanza of the Pacific North-west. While stationed in the West Indies, he collected plants for Joseph Banks and later became one of Kew's plant hunters. On his second expedition to the Americas, Menzies discovered many fine plants. Unfortunately he was confined to his cabin in the ship for the last three months of the trip by the belligerent Captain Vancouver who objected to his crew helping Menzies with his plant collecting. This meant that he was able to present Banks with only two plants when he returned to Kew, grown from some strange nuts served for dessert by the Spanish Viceroy at Valparaiso in Chile. This was Menzies's finest introduction – *Araucaria araucana* or the monkey puzzle (see page 137). Menzies's bad luck was a

fellow-Scot's good fortune, and it fell to David Douglas to reveal the wonder of this region's flora. Although credited with introducing over 200 new species, it is for his lofty conifers, which changed the garden landscapes of Britain,

A close-up of the needles and cones of the Douglas fir (*Pseudotsuga menziesii*). This was the tallest tree in the world until extensive logging destroyed many thousands of acres of forest.

Europe and America, that Douglas is primarily remembered. One still commemorates his name – the Douglas fir (*Pseudotsuga menziesii*), one of Menzies's 'lost' finds.

Douglas was an enigmatic figure, and his character continues to intrigue biographers. Born in Scone, near Perth, on 25 July 1799, he was a spirited child who developed twin passions for the outdoors and natural history. In 1810, at the age of eleven, he began a seven-year apprenticeship to become a gardener, under the watchful eye of William Beatty, head gardener to the Earl of Mansfield at Scone Palace. Douglas continually surprised and impressed Beatty with his eagerness and ability to learn. He persuaded Beatty's assistant, Mr McGillivray, to teach him elementary botany and was captivated by tales told to him by the Brown brothers, local nurserymen, of botanizing in the Highlands. After spending a winter improving his mathematics and sciences, Douglas moved to the gardens at Valleyfield, near Culross in Fife. Here his inquiring mind and determination came to the attention of the owner, Sir Robert Preston, who gave him a free run of his extensive library. Douglas took full advantage of this unusual privilege, and two years later he secured a position at the Botanic Gardens in Glasgow.

April 1820 was a red-letter month for botany in Glasgow. Not only did Douglas begin his new job, but William Hooker arrived at the university to take up his seat as Professor of Botany. This good-natured man quickly became one of the university's most popular lecturers, and in almost a dummy run for his later career as the first official Director of Kew, displayed his organizational skills and boundless energy revitalizing the Botanic Gardens. In the course of the shake-up he became aware of the new, diligent junior member of staff. The two became firm friends, and Hooker taught Douglas much about practical botany as they roamed the Scottish hills in search of material for Hooker's *Flora Scotica*. Douglas was a frequent visitor to Hooker's home, and in later years he was an inspiration to William's younger son, Joseph, with his tales of strange, far-off places, much as the Brown brothers had been to him. He even taught Joseph how to fish like a Native American in a pool in the back garden.

Douglas's big break came in the spring of 1823 when Joseph Sabine, the Secretary and driving force behind the Horticultural Society of London (now the Royal Horticultural Society), asked Hooker to recommend him a suitable botanical collector to work for the Society. Hooker without hesitation put forward the twenty-four-year-old Douglas's name. The Horticultural Society had been formed in March 1804, when a group of eminent horticulturists, both amateur and professional, got together to form a Society 'whose object should be the improvement of Horticulture in all its branches'. Among the activists who met above Hatchard's book shop were John Wedgwood, son of the famous potter and future uncle of Charles Darwin, and Sir Joseph Banks. As Kew entered a period of decline, the Horticultural Society went from strength to strength. In 1823 33 acres of Chiswick were rented from the Duke of Devonshire in order to develop experimental gardens, and the Society took up the mantle of Kew and began to send out its own plant hunters: in 1821 George Don was dispatched to West Africa and South America, while John Potts went to China and the East Indies; in 1822 John Forbes went to East Africa and a second China expedition was undertaken by John Parks in 1823.

When David Douglas travelled to London from Glasgow he expected to be sent out to China. Unfortunately, political unrest there forced Sabine to change his plans, and Douglas, somewhat disappointed, was packed off to New England, on the east coast of North America. This was the first of many instances of bad luck which dogged his plant-hunting career. He reached New York on 5 August 1823 after an appalling crossing during which he had endured bad weather and a shortage of food, and when he came to disembark he was told by immigration officials that he was too scruffy, and permission to land was refused until he had purchased new clothes.

Once ashore, the now dapper Douglas set out on a four-month journey, exploring the New York vegetable markets and the flower garden of Mr Van Ransalier in Albany before heading for Buffalo and Amherstburg on Lake Erie. Here he first encountered truly wild America, and perhaps with memories of his childhood flooding back he fell under its spell. He collected seed from veronicas, eupatoriums, helianthemums, liatris, solidago and asters, and used his gun to shoot branches from a tall oak for the leaves and acorns. Wherever Douglas went misfortune was not far behind, and on this journey it caught up three times. On the first occasion he was out riding when his horse bolted. Holding on for dear life, he kept shouting commands to the horse, only to find when at last the beast stopped that it understood only French! On the second, having entrusted his possessions and money to his cart driver while he climbed a tall tree to collect seed, he was surprised to see his companion running off into the woods with all his belongings. Douglas could not drive the cart and was left stranded and penniless. Finally, on the trip back to Buffalo his boat almost sank in a violent storm. The philosophical Douglas continued his journey via Niagara Falls (where he found an astragalus and a viola) to

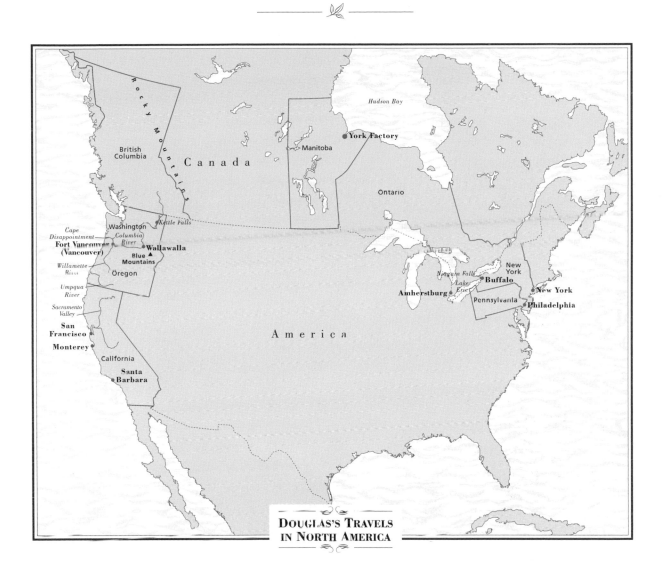

DOUGLAS'S TRAVELS
IN NORTH AMERICA

Philadelphia, then returned to New York, stopping to collect saracenias in a swamp on the edge of the Hudson river. He set sail for Britain, arriving home on 10 January 1824. When he revealed the extent of his collection, the trip was heralded by the Horticultural Society as 'a success beyond our expectations'. Among his trophies were several new ornamental species and a wide selection of apple, pear, plum, peach and grape varieties. The latter had been bred in America and found a warm welcome in the kitchen gardens of Britain's country houses.

Douglas's next collecting trip for the Horticultural Society was more extensive. This time he was travelling to the largely unexplored wilds of the Pacific

North-west, and he left Gravesend aboard the Hudson Bay Company ship *William and Ann* on 26 July 1824. The voyage via Cape Horn to the mouth of the Columbia river on the Washington/Oregon state borderline took a tedious eight and a half months, with stop-overs at Madeira, Rio de Janeiro, Juan Fernandez and the Galapagos Islands, where he was able to explore and study the flora. He arrived at the uninvitingly named Cape Disappointment on 9 April 1825 to find that the weather alternated between thick mist and heavy rain. His spirits were raised, however, when he saw the scenery:

> With respect to the appearance of the country and its fertility my expectations were fully realised. It is very varied, diversified by hills and extensive plains, generally good soil. The greater part of the whole country as far as the eye can reach is closely covered with pine of several species. In forest trees there is no variety or comparison to the Atlantic side, no beech, gleditschia, magnolia, walnut, one oak, one ash. The country to the northward near the ocean is hilly … on the south side of the river [it] is low and many places swampy.

And it was here that he made his first floral discovery: 'On stepping on the shore *Gaultheria shallon* was the first plant in my hands. So pleased was I that I could scarcely see anything but it.'

Here, at its mouth, the Columbia river measured some 5 miles across, but heading inland it ran deep and swift. It was a rich source of salmon and sturgeon and acted as the principal transport route for both Native Americans and pioneers. Douglas initially viewed the indigenous population with some suspicion and trepidation, but as he became accustomed to their ways he grew to admire them, while remaining on his guard – many tribes were understandably hostile to European settlers. The Hudson Bay Company (which had promised to help him) was in the process of moving its base, Fort Vancouver, from the coastal region to a site some 90 miles upstream, and Douglas was greeted warmly by the chief factor of the camp, John McLoughlin. He made the fort his base for the next two years. At first, he lived in a tent among the log cabins but, as his plant collection grew, he moved into a deerskin lodge and finally a bark hut. He soon became used to the rough lifestyle:

> In England people shudder at the idea of sleeping with a window open; here, each individual takes his blanket and with all the complacency of mind that can be imagined throws himself on the sand or under a bush just as if he was going to bed. I confess, at first, although I always stood it well and never felt any bad effects from it, it was looked on by me with a sort of dread. Now I am well accustomed to it, so much so that comfort seems superfluity.

A sketch of Fort Vancouver, which Douglas made his base throughout
his time in the Pacific North-west.

Douglas spent much of his first summer making local trips to familiarize
himself with the flora and fauna. It was on one of these that he discovered
that favourite of the spring garden, the Oregon grape (*Mahonia aquifolium*),
and the conifer that bears his name, which 'exceeds all trees in magnitude.
I measured one lying on the shore of the river 39 feet in circumference and
159 feet long; the top was wanting … so I judge that it would be in all about
190 feet high … they grow very straight; the wood is softer than most … and
easily split.'

At first Douglas accompanied white traders and trappers, but he soon built
up a rapport with the Native Americans and began to use them as guides. He
noted with wry amusement that his plant-collecting activities earned him the
name of 'Grass Man', and one tribe deemed him to be an evil spirit when they
saw him drinking an effervescent health drink – they thought it was boiling
water! He also found that lighting his pipe with a magnifying glass and simply
wearing a pair of spectacles provoked displays of incredulity (quite possibly
feigned by the Native Americans for their own amusement).

In May Douglas found the popular flowering currant (*Ribes sanguineum*), *Lupinus polyphyllus* and *Clarkia pulchella*, and on 20 June he embarked in a precarious canoe to make a month-and-a-half-long journey up the Columbia river to the Grand Rapids and Great Falls. Progress was slow, and was made especially tiring by the constant battle against fast-flowing waters and contrary winds. On the positive side, the enforced delays while waiting for the wind to drop allowed him to make a number of excursions into the neighbouring countryside. The terrain was at times extremely challenging and the ambitious plant hunter often pushed himself to the limits of his endurance:

> The luxury of a night's sleep on a bed of pine branches can only be appreciated by those who have experienced a route over a barren plain, scorched by the sun, or fatigued by groping their way through a thick forest, crossing gullies, dead wood, lakes, stones, &c. Indeed so much worn out was I three times by fatigue and hunger that twice I crawled, for I could hardly walk, to a small abandoned hut. I had in my knapsack one biscuit …

On the third occasion Douglas shot two partridges but fell asleep while cooking them. He awoke to find that his dinner was burnt to a cinder and his cooking pot ruined. Undaunted, he got up and, using the lid of his tinder-box as a makeshift kettle, brewed himself a cup of tea 'which is the monarch of all food after fatiguing journeys'. No stranger to adversity, he seems to have thrived in such harsh conditions, his austere Scottish roots standing him in good stead.

On his way back to Fort Vancouver Douglas visited the local chief of the Chenook and Chochalii tribes, Cockqua, who as a mark of friendship cooked a giant sturgeon measuring 10 feet long and weighing 400–500 lb. As guest of honour Douglas was given the choicest parts of the fish – the head and spine. Cockqua was at war at the time with the Cladsap tribe, which lived across the river, and he entreated Douglas to sleep in his hut for safety. He refused, thinking it would smack of cowardice, and spent the night in his tent some fifty yards from the village. In the morning he was lauded for his display of bravery by many of the Native Americans, but one young brave was clearly irked by the praise. He gave a display of his skill with bow and rifle, firing arrows through a 6-inch metal hoop thrown in the air and shooting to with an inch of a target at 110 yards. Douglas, who was carrying a shotgun, responded by disturbing a large eagle perched nearby and shooting the poor bird down in mid-flight. Unimpressed, the Native American threw his hat in the air and challenged the

Although many of Douglas's major discoveries were conifers, he also found
several lovely shrubs. The flowering currant (*Ribes sanguineum*) has become a favourite
in the spring garden.

white man to hit it. Douglas rather gleefully records that he 'carried the whole
of the crown away, leaving only the brim'.

When he returned to Fort Vancouver on 5 August, Douglas heard that a ship
was to leave for Britain within the next few weeks. After a brief sortie up the
Multnomah river on the 19th, where he first became aware of the sugar pine
('In the tobacco pouches of the Indians I found the seeds of a remarkably large
pine which they eat as nuts, and from whom I learned it existed in the mountains to the south. No time was lost in ascertaining the existence of this truly
grand tree, which I named *Pinus lambertiana*'), Douglas busied himself packing his botanical collection and made preparations for the journey to the
mouth of the Columbia river. He arrived an hour after the ship had set sail,

however, after cutting himself on a rusty nail and developing an abscess in his knee joint. Cursing his bad luck, he turned despondently north for a brief trip along the Chehalis river and on to Cowlitz river. It rained continuously, and when the food rations ran precariously low he was forced to subsist on a diet of roots and berries. By the time he got back to the comfort of his little bark hut at the end of 1825, Douglas was due a much needed rest. Since his arrival eight months earlier he had travelled 2,105 miles by foot, horse and canoe, and 1826 would prove to be even more arduous – he would almost double the distance covered in the previous year, travelling a staggering 3,932 miles.

The winter weather made travelling unfeasible, and as it was the wrong season for plant hunting, Douglas holed up in Fort Vancouver until the spring of 1826. His first trip of the year was eastward to the Spokane river and the Kettle Falls. He set off on 20 March, taking with him almost no clothes but two reams of drying paper (for making herbarium specimens) – his dedication to science proved stronger than his need for creature comforts. It was a predictably tough journey and lasted until late August. On 10 May the party had to ford the icy Barriere river and, finding no local transport, they decided to swim across rather than waste time building a raft. Carrying his personal possessions above his head, the naked Douglas swam across the river twice. The clouds began to darken and soon a heavy hailstorm assailed the travellers. After spending thirty minutes in the freezing water the numb botanist had to kindle a fire to revive himself and his guides. Perhaps at this juncture he glumly regretted the warm, dry clothes he had left behind. His sacrifice was not in vain, however, for on this trip he acquired the lovely *Lilium pandicum*, the graceful Western yellow pine (*Pinus ponderosa*), the 'exceedingly beautiful' *Erythronium grandiflorum* and the cheerful *Penstemon glaber*.

In June Douglas moved on to Wallawalla, from where he made two excursions into the Blue Mountains. Here he was plagued both by mosquitoes and by bad weather, but as usual these discomforts were stoically (or masochistically) endured. No European had scaled the heights of the Blue Mountains, and Douglas's determination to be the first white man to conquer the towering peaks often put him in danger. After ploughing through the snowdrifts in the foothills of one of the highest peaks, Douglas found the rest of the ascent fairly easy going. Once at the summit he became mesmerized by the view, but:

> I had not been there above three-quarters of an hour when the upper part of the mountain was suddenly enveloped in dense black cloud; then commenced a most dreadful storm of thunder, lightning, hail, and wind. I never beheld anything that

A sketch by Sir George Back, made in 1833, showing men carrying their canoe
across the Hoarfrost river. Douglas had many similar experiences during his two years
of explorations around Fort Vancouver.

could equal the lightning. Sometimes it would appear in massy sheets, as if the heavens were in a blaze; at others, in vivid zigzag flashes at short intervals with the thunder resounding through the valleys below, and before the echo of the former peal died away the succeeding was begun, so that it was impressed on my mind as if only one. The wind was whistling through the low stunted dead pines accompanied by the merciless cutting hail.

Douglas hastily headed back to his camp, which he managed to reach just as last light was fading. His tatty clothes were soaked, so he undressed and wrapped himself in his blanket. At midnight he was woken up by the cold and when he tried to get up discovered that his knees 'refused to do their office'.

Douglas found many fine new species on his excursions into the Blue Mountains, including *Paeonia brownii* (the only North American paeony), *Lupinaster macrolephalus*, *Trifolium altissimum* and the golden-flowered *Lupinus argenteus*, as well as several new phlox. Trouble still caught up with him once in a while. On one particularly nerve-racking occasion Douglas found himself taking on the role of diplomat when a furious row broke out between a Native American chief and an interpreter. The chief accused the 'man of language' of not translating faithfully, and in the ensuing tussle the poor interpreter had a handful of hair pulled out. Enraged by the encounter, the chief returned with a party of seventy-three armed warriors. The dispute was eventually resolved by an exchange of presents and no blood was spilt.

During this time, Douglas began to experience trouble with his eyesight – the blowing sands of the desert-like country around Wallawalla and minor snow blindness contracted in the Blue Mountains combined to cause painful inflammation. Over the following years his sight would continue to deteriorate and may have contributed to his gruesome end. Although Douglas took hardship in his stride, he was not impervious to the melancholy induced by prolonged isolation. He desperately looked forward to mail deliveries from England and recorded in his journal that he would get up in the middle of the night several times to read newly received letters from home. Once again he heard that a boat was leaving for England and, determined this time to see his precious collection shipped home, he hurried back to Fort Vancouver. His bedraggled appearance as he entered the fort on 29 August caused grave concern among the inhabitants, who assumed that he was the sole survivor of some terrible disaster. Barefoot and dressed only in a tattered shirt, ripped trousers and an old straw hat, Douglas had to admit that 'my careworn visage had some appearance of escaping from the gates of death'.

Douglas had already seen seed of the sugar pine and a French hunter had presented him with a cone measuring over 16 inches in length. But he had yet to see this giant, and when he heard that a party led by A. R. McLeod was to leave for the headwaters of the Willamette river, he joined the group. Setting off on 20 September, they headed south on a journey that was as miserable and uncomfortable as any Douglas had so far experienced. From mid-October it rained constantly and there was insufficient food. The countryside was forested and mountainous, making progress frustratingly slow. Yet despite the dreadful conditions Douglas managed to amass a fine collection of seeds as he walked along, including the madrona (*Arbutus menziesii*) and the headache tree (*Umbellularia californica*). The latter is named because if the leaves are crushed and sniffed, one develops a splitting headache lasting about half an hour. Even Douglas observed that 'so exceedingly powerful is the fragrant scent which it emits by the rustling of its leaves that it produces sneezing'. He also noted that a 'decoction' made from the bark was used by trappers to make a beverage.

Douglas reached the Umpqua river on 16 October and after an enforced week's rest (he had injured himself falling down a ravine), he set out for sugar pine territory. On the night of 24 October he was caught up in yet another horrific storm. His tent blew down, and as it was impossible to light a fire he spent a wretchedly cold night wrapped in a blanket and the canvas of the tent:

> Sleep of course was not to be had, every ten or fifteen minutes immense trees falling producing a crash as if the earth was cleaving asunder, which with the thunder peal before the echo of the former died away, and the lightning in zigzag and forked flashes, had on my mind a sensation more than I can ever give vent to; and more so, when I think of the place and my circumstances. My poor horses were unable to endure the violence of the storm without craving of me protection, which they did by hanging their heads over me and neighing.

When finally the storm blew itself out, Douglas tried to bring some life into his frozen frame by rubbing himself down with a handkerchief in front of a fire which had at last been able to be lit. Fortunately his perseverance was to reap rich rewards, for next day he 'reached my long-wished *Pinus* ... and lost no time in examining and endeavouring to collect specimens and seeds'. One fallen specimen measured 215 feet in length. As the cones on the living trees grew beyond climbing height, Douglas began firing his gun into the branches to dislodge some. A party of eight Native Americans, drawn by the sound of gunfire, entered the clearing and approached him. They were armed with bows, bone spears and flint knives and appeared 'anything but friendly'. Douglas attempted to explain what he was doing and, apparently satisfied, the braves sat down to smoke. However, one of the group began to string his bow and another sat sharpening his knife. Unable to escape, Douglas cocked his gun and drew one of his pistols. A stand-off lasting a nerve-racking ten minutes ensued, until the leader demanded some tobacco. Douglas replied that they would be welcome to some if they helped him collect pine cones. As soon as the Native Americans had left the clearing Douglas picked up his belongings and ran back to his camp, where he spent the night listening for their return. To add to his worries, a grizzly bear had attacked his guide earlier in the evening and was roaming nearby. As dawn broke, the bear shambled into the camp with her two cubs. Fearing for his safety, Douglas rode his horse to within 20 yards of the bears before shooting the mother and one of the cubs.

Douglas returned for a much needed winter of recuperation to Fort Vancouver, where he spent the harsh months preparing his collection and visiting his friend Cockqua. It has been suggested that it was in fact the charms of Cockqua's pretty young daughter that repeatedly enticed Douglas back to the village near Gray's Harbour, but the upright Scot was too discreet to record such details. As the plan was to publish an account of the expedition, he would not have wanted to give any grounds for salacious gossip.

The following spring Douglas joined the annual Hudson Bay Company Express on its transcontinental trek of 995 miles. Departing on 20 March 1827, the energetic plant hunter walked for the first twenty-five days but developed sore feet and had to transfer to canoe. At the end of April the party reached the Rockies, where Douglas was greatly impressed by the rugged beauty of the mountains, although his enjoyment of the scenery was somewhat marred by the conditions underfoot. He had never mastered the art of walking in snowshoes and frequently had to extricate himself from snowdrifts. Fortunately, there was time for a few minor diversions, and Douglas climbed one of the mountains beside the trail. He christened it Mt Brown, and named its lofty neighbour after his old friend Sir William Hooker. Once clear of the Rockies the express pushed itself hard, often covering over 40 miles in a single day. Tragedy nearly struck in May when one of the party, Mr F. McDonald, was seriously injured during a buffalo hunt when a wounded bull tossed him into the air and gored him seven times. A shot from one of the party startled the beast and, after gently nudging the inert man over, it trotted off. Douglas administered some vital first aid and despite horrific wounds McDonald eventually made a full recovery.

For once Douglas completed a journey without serious mishap to himself, although he was upset by the death of his pet Calumet eagle, which had strangled itself on one of its restraining cords. Having covered a total of 7,032 miles in the past three years, he safely reached York Factory on Hudson Bay on 28 August 1827. But his adventures in North America were not quite over. He decided to pay a visit to the *Prince of Wales*, on which he had booked a passage home, and rowed out to it with two companions. On the return trip a tremendous storm erupted, blowing the little rowing boat 70 miles out into the bay, and it was only with the greatest of good luck and a favourable current that they managed to limp back to shore.

Douglas understandably took a considerable time to recover from this last ordeal, but his triumphant return to England on 11 October 1827 must have gone some way towards healing the physical and mental scars. Unfortunately, the adulation he received from high society seems to have turned the taciturn Scot into a garrulous show-off. He argued with friends and associates, including Sabine, and became bored and restless with life. He attempted to write up his adventures but found that his literary skills did not match up to his botanical ones. It was only the direct intervention of his friend and mentor Sir William Hooker that saved his career. Douglas was granted his wish to return to North America, and he left Portsmouth on 26 October 1829, almost two years to the day since his return.

During his last expedition Douglas flitted about the west coast of America. He revisited some of his plant hunting grounds around the Columbia river during the summer of 1830. Near the Columbia Cascades he gathered seed from the giant fir (*Abies grandis*) and the red silver fir (*Abies amabilis*), trees that he had first discovered in 1825 but had not introduced. Trouble between Native Americans and the settlers made travelling dangerous, so Douglas sent back three chest-loads of seed in October before sailing south to San Francisco, arriving there in early 1831. From here he headed south to Monterey in California, where he found the Monterey pine (*Pinus radiata*), and on to Santa Barbara before turning northwards towards the Sacramento Valley. Although he aimed to return to the Columbia river in the summer of

---

Overleaf: The spectacular landscape of the Pacific North-west enraptured Douglas, but he endured much hardship in his quest for the 200 or so species he introduced.

1831, there was no shipping available and he was forced to stay another season in California. Perhaps this was a blessing in disguise, for he discovered a wealth of new plants in the warmer climate of California. He wrote to Sir William Hooker saying he had found 360 new species and twenty new genera, including the noble fir (*Abies procera*). In another letter Douglas suggests that he has also re-found Menzies's coastal redwood (*Sequoia sempervirens*), describing the trees as 'some few upward of 300ft' and stating he has procured 'fine specimens and seed also'. What happened to the seed is a mystery; it has been suggested that it was lost when his canoe capsized in June 1833, but this is not feasible because Douglas dispatched his Californian collection, some 670 plant species, to England in August 1832 before sailing to the Columbia river.

After a brief stop-over in Hawaii, Douglas reached Fort Vancouver on 14 October. It is clear from his letters that he was keen to visit Alaska and make his way home via Siberia (a feat that only a man of Douglas's calibre could have even considered, let alone achieved). After a few local expeditions and over-wintering in Fort Vancouver, he set off on 19 March 1833 on his most ambitious journey. He managed to reach Fort St James, where he turned back for reasons unknown. His canoe ran on to rocks in the Frazier river and he was thrown into the turbulent water, dragged into a whirlpool and spun around for more than an hour. Douglas was lucky to escape with his life, but he lost all his personal possessions, his plant collection and his journal. Downcast and suffering from poor health (he had now lost the sight in one eye), he turned his attention back to the alluring islands of Hawaii and took a passage from San Francisco aboard the *Dryad*, arriving on 23 December 1833.

In Hawaii Douglas climbed Mauna Kea and Mauna Loa and saw the active volcano Kilaeua. In early July 1834 he set off again to walk across Mauna Kea, accompanied by his faithful little Scots terrier Billy and a servant, John. At some point during the walk the two men parted company and Douglas continued alone. On the morning of the 12th he called at the hut of Ned Gurney, an ex-Botany Bay convict who trapped the island's wild cattle for a living. Gurney employed a devious trapping technique – he dug a deep pit on the side of a water source, covered it with a flimsy roof and surrounded the area with a stockade. The cattle fell in while trying to get to the water. After breakfast Gurney accompanied Douglas along the path for a mile or so and warned him about the pits. What happened next is a matter of conjecture. The most likely explanation is that on hearing an animal trapped in one of the pits Douglas went over to investigate, lost his footing and fell. His gored and trampled body was discovered later in the day by some passing locals. Rumours were spread that he had had an affair with Gurney's wife and that in a fit of jealous rage the trapper had pushed the Scot into the pit, but no evidence was ever found to indicate foul play. A post mortem carried out by four doctors at Honolulu confirmed that the bullock had inflicted the fatal injuries.

Douglas is credited with introducing over 200 new species, including spectacular conifers that changed the face of the Victorian landscape, but his contribution is often overlooked. Few people realize that Britain has only three native conifers (the yew, Scots pine and common juniper) and that the towering conifers which grace the skyline of so many designed landscapes are in a large part the result of Douglas's ceaseless wanderings through a distant and often inhospitable land. His introductions also form the basis of the forestry industry in North America (Douglas fir), Britain (Sitka spruce) and many southern hemisphere countries (Monterey pine).

Douglas's trips were made on behalf of the Horticultural Society and were paid for by subscriptions from some of its wealthy members, who in return received seed of the new, exotic and rare plants found on the expedition. This policy of allocating the new plants to a privileged few in exchange for funds to send out plant hunters worked very well for several reasons. First, it meant that the society was at the forefront of introducing new plants and its reputation (as well as its garden at Chiswick) blossomed. Second, the patrons were kept happy as the influx of plants fed their insatiable appetite for novelty, while being seen to be growing rare species reflected their superior social status. Third, by operating in this way the Society ostensibly avoided competition with commercial nurserymen, but since several of these were also patrons, the new plants did find their way into the general marketplace.

By the 1830s the enormous cost of the Napoleonic Wars had been recouped and trade, both internal and foreign, was booming. For entrepreneurial businessmen and merchants these were prosperous times, and one way to display a newly made fortune was to acquire a country estate. Industrialization went hand-in-hand with urbanization, and the suburban villa with its garden of anything between 1 and 10 acres became *de rigueur* for the prosperous.

A few of these *nouveaux riches* employed the services of a professional designer, but most turned to the ready-made gardening advice provided by the influential garden-writer John Claudius Loudon (1783–1843). Loudon was the first to exploit this large new market and distributed easily digestible horticultural information through his books and his new concept, the gardening magazine. He published his *Encyclopedia of Gardening*, an epic of over 1,000 pages, in 1822 and began *The Gardener's Magazine* in 1826.

Loudon was also responsible for a new approach to garden-making that ideally suited many of Douglas's new conifers. The English Landscape School had focused on imitating and improving nature to a point where it became impossible to tell where the designed landscape ended and the wild, untamed countryside began. Loudon now suggested that the garden should be 'an imitation of nature, subject to a certain degree of cultivation or improvement, suitable to the wants and wishes of man'. This concept, termed the 'Gardenesque', was widely misinterpreted by the eager new gardeners, who decided that the garden should be a work of art rather than an imitation of nature. The layout of the garden changed and the stamp of man became clearly visible. Around the house, instead of the sea of grass advocated by 'Capability' Brown, the architectural terrace made a comeback. This was planned with geometric flowerbeds planted with predominantly exotic foreign species. In these beds Douglas's hardy annuals found a transitory place but were superseded when the brilliantly flowered tender annuals arrived from South America and Africa. These went on to become the staples for the often garish summer bedding displays that were to be so fashionable in Victorian gardens.

Douglas's highly decorative and ornamental conifers, with their wide range of shape, size, colour and unusual needles and cones, were perfectly suited to display the dominance of art, but in another way. They were carefully arranged to best show their character and beauty and were set within the manicured lawns and beds that constituted the pleasure grounds beyond the formal garden. This arrangement not only emphasized man's ingenuity in creating artistic planting arrangements but also showed that man cultivated the garden rather than leaving it to the untidy vagaries of nature. The new trees were particularly favoured because of their varied evergreen foliage. This made possible the design of displays that could be admired throughout winter, something that had previously been difficult to achieve. Individual specimens or groups, or both, were used to provide a focus of attention or were positioned to create a permanent frame for a view or vista.

By the 1830s theme gardens or gardens-within-a-garden had become fashionable. One very popular theme was to make a botanical or geographical plant collection, a favourite being the 'American Garden'. Tracing its origins back to the late eighteenth century, this was originally, as its name implies, a garden compartment planted with the recently introduced North American and

Canadian species. Douglas's new North American conifers fitted both bills and perfectly suited the Victorian collector's mentality. Not only were they included in the American garden but they greatly augmented the theme garden that catered specifically for evergreens – the pinetum. Dating back to the eighteenth century (Kew's was begun in 1760), the pinetum was, in its most literal sense, 'a complete collection of all the Coniferous trees and shrubs known', but in reality the range of species grown depended on interest, the size of the garden and the owner's wealth. Exotic conifers were not new to Britain – the Tradescant introductions had been complemented by the cedar of Lebanon (*Cedrus libani*), introduced from Asia Minor in about 1645, the maidenhair (*Ginkgo biloba*) from China in *c*. 1758, and William Kerr's later Chinese collection of 1804 – but it was Douglas's introductions that ignited the Victorian passion for them, a passion that was to grow increasingly intense as more new conifers were brought back from North and South America, the Himalayas and the Far East.

Conifers were used in several other ways. They could be mixed with deciduous trees, either planted among existing woodland to increase the species diversity, or included in an arboretum, essentially a collection of trees, native and exotic, deciduous and evergreen, grouped together for aesthetic and botanical interest. They could be planted in an avenue, a feature that took a bewildering variety of forms within the landscape, in addition to lining the drive to the house. As with so many Victorian artistic concepts, however, enthusiasm outpaced good taste. A more utilitarian use of Douglas's conifers was the perimeter shelterbelt. Many exposed gardens were surrounded by evergreen trees to provide all-year-round wind protection, thus creating a more conducive microclimate within the garden. In coastal locations, such as the south coast of Cornwall, the Monterey pine was particularly favoured for its quick growth and tolerance to salty winds.

David Douglas was the first large-scale privately sponsored explorer sent out with the express purpose of finding exciting new garden plants. In this he succeeded beyond expectation, and as a result of his work (and that of later collectors), conifers became an integral part of the Victorian landscape. In the next chapter we return to scientific collecting for Kew, but also follow the journey of a man whose introductions had an even greater impact on the garden.

David Douglas's Plant Introductions

The date beside each plant name is the date of its introduction into Britain.

Lupinus polyphyllus (1826)

Lupinus = lupin; from (Lat.) *lupus*, the wolf, for the erroneous belief that these plants destroyed soil fertility

polyphyllus = (Gk) with many leaves

Whorled spires of gorgeous satin-blue and reddish-purple flowers 6–24in (15–60cm) long are held on 4ft (1.5m) stems from May to July. This vigorous perennial possesses compound palmate leaves with 9–17 leaflets and is an ancestor of the popular garden lupins. Hybridizing with additional species, e.g. *L. hartwegii*, a magenta-flowered annual, and *L. arboreus*, the yellow-flowered tree lupin, has produced a broad colour range.

Grows in wet, grassy meadows from San Francisco, California, to British Columbia.

Ribes sanguineum (1826)

Ribes = from the Arabic *ribas*, acid-tasting

sanguineum = (Lat.) blood-red

Drooping racemes of deep pink flowers festoon the branches of the flowering currant in April. A rounded medium-sized deciduous shrub with pungent, five-lobed green leaves, multi-coloured in autumn. Good cultivars include 'King Edward VII' (crimson), 'Pulborough Scarlet' (deep red), and 'Brocklebankii', with golden yellow foliage. Douglas also introduced *R. speciosum* (1828), which has beautiful red fuchsia-like flowers.

Ranges from northern California to British Columbia in conifer forest at 1,950–5,700ft (600–1800m), growing from 3–10ft (1–3m). Originally discovered by Archibald Menzies in 1793.

Pseudotsuga menziesii (1827)

Pseudotsuga = (Lat.) false hemlock

menziesii = after plant hunter Archibald Menzies

Sky-scraping corky-barked trunks bear down-sweeping branches densely clothed in ranks of 1¼–4in (3–10cm) deep green leaves, emitting a strong fruity smell when crushed. The Douglas fir is an impressive, fast-growing tree.

Grows in dense forest stands from south-west British Columbia to central California, where it reaches up to 310ft (95m) with record-breaking specimens of up to 387ft (118m) tall grown in the last century.

Garrya elliptica (1828)

Garrya = after Nicholas Garry

elliptica = (Lat.) elliptic; longer than wide, broadest in the middle, with curved sides

Cascades of 6in (15cm) long soft grey-green and purple-brown catkins adorn this superb winter-flowering shrub from January to March. The leathery evergreen leaves are grey-green and woolly beneath. Male plants carry the finest catkins and female plants pendulous clusters of purple-brown fruits. The selection 'James Roof' has catkins up to 14in (35cm) long.

Occurs in forest and brushwood up to 1,950ft (600m) in the coastal ranges of southern California to Oregon, reaching 26ft (8m) tall.

Abies grandis (1830)

Abies = (Lat.) fir

grandis = (Lat.) big, showy

Towering, columnar trees are clothed in fragrant, dark green 1–2in (2–5cm) leaves, marked on the underside with two greenish-white bands, and emitting a strong aroma of tangerines when they are crushed. The fast-growing grand fir is the world's tallest fir and when mature bears bright green 3–4in (7–10cm) cylindrical cones, which ripen to dark brown.

Ranging from Vancouver Island and British Columbia, south to Caspar in California and east to Idaho, this tree grows up to 245ft (75m), possibly up to 295ft (90m), in height and is widely grown for timber.

Abies procera (1830)
Abies = (Lat.) fir
procera = (Lat.) tall

Bright blue leaves in upward sweeping combs swathe the magnificent silver-barked noble fir. A large conical to broadly columnar tree with interesting raspberry-like male flowers in May (even on quite young trees), followed by huge 8–10in (20– 25cm) female cones (whose weight can break branches), which turn soft golden-brown and have whorls of decorative downward-pointing green bracts.

Occurring on the Cascade Mountains and some peaks of the Coast Range of Oregon and Washington, from 1,950–5,000ft (600–1500m), growing from 140–190ft (45–60m). Grown elsewhere for its timber.

Pinus radiata (1833)
Pinus = (Lat.) pine
radiata = (Lat.) of radiating form

Dome-shaped canopies of densely packed 6in (15cm) long, deep green leaves carry whorls of huge 6in (15cm) cones atop bare trunks of deeply fissured, dark brown bark. The Monterey pine is an impressive, fast-growing large tree with yellow male cones in spring. The female cones may remain intact on the tree for up to thirty years or more.

Confined to three localities on the Californian Monterey peninsula and two Mexican islands, it reaches only 65ft (20m) in the wild. However, it has been extensively planted worldwide as a timber tree and windbreak in coastal areas, attaining incredible improved growth rates in its new range, e.g. 213ft (65m) in forty years in New Zealand.

Garrya elliptica is an attractive evergreen shrub, which enlivens the winter garden, first with its catkins, and later with its purple-brown fruits on the female plant.

4

On Top of the World

Sir Joseph Dalton Hooker

(1817–1911)

JOSEPH DALTON HOOKER HAS BECOME RECOGNIZED AS THE MOST important botanist of the nineteenth century and one of the key scientists of his age. This remarkable man, a close friend of Charles Darwin, was a prolific author, the second director of the Royal Botanical Gardens, Kew, and responsible for initiating rhododendromania.

Born in Halesworth, Suffolk, on 30 June 1817, the second son of Sir William Hooker, Joseph spent his childhood in Glasgow, where his greatest joy was to help his father with his herbarium. Sir William nurtured his son's keen interest in plants and allowed his son to accompany him to his college lecture every morning. Later in his life Joseph fondly recalled an incident in his childhood

Left: Sikkim, a wild, mountainous and inhospitable country, squeezed between Nepal, Tibet and Bhutan

Above: Joseph Hooker at the age of twenty-two. From a sketch by William Taylor, 1839.

when, at the tender age of five or six, he was discovered 'grubbing in a wall in the dirty suburbs of the dirty city of Glasgow'. When asked what he was doing, he proudly announced that he had found the moss *Bryum argenteum*. His passion grew, and by the age of ten even his modest father boasted of his son's enormous knowledge. It was at this time that Joseph developed his other lifelong love – that of exploration. He spent many hours reading Mungo Park's *Travels* in search of the source of the Niger and Captain Cook's *Voyages*, and dreamed of retracing their steps. He was further inspired to become an explorer by David Douglas's tales of his adventures in the wilds of North America.

Joseph attended Glasgow High School, where he benefited from the liberal Scottish education system, and at the tender age of fifteen entered Glasgow University to study medicine. It was while he was at university that two meetings took place that changed his life: a chance encounter with Charles Darwin in Trafalgar Square began a close friendship, and in September 1838 he breakfasted with Captain James Clark Ross. Ross was about to lead a British Association expedition to the Antarctic, and Joseph determined to join. With his father's help, the twenty-two-year-old secured the position of assistant ship's doctor and botanist, and on 28 September 1839 he sailed out of the Medway on board Her Majesty's discovery ship *Erebus*. Ross, who was in command of both *Erebus* and her sistership *Terror*, was charged with the task of pinpointing the position of the magnetic south pole, and his previous experience of the treacherous Arctic waters was to prove invaluable. On the night of 13 March 1842 a severe gale blew up as the ships were passing through an ice field and as a particularly large iceberg loomed ahead, the *Terror* crashed into the *Erebus*, entangling the ships. Eventually they broke free, but the *Erebus* was left severely damaged and facing the oncoming iceberg. Ross's adept captaincy saved the ship. He quickly gave the order to about stern and steered the *Erebus* between two walls of ice. Despite such dangers, the expedition was a complete success and even set a new world record when they penetrated deep into the Antarctic circle and crossed the 78° parallel.

During the four-year trip Hooker was able to botanize extensively on three continents as the ships visited Madeira, the Cape of Good Hope, Tasmania, New Zealand, Australia, the Falkland Islands and the southern tip of South America. He carefully recorded the exact locations of new plant discoveries and observed the similarities with other species from different locations. This laid the foundations for his authority on the geographical distribution of plants, which was to prove vital to Darwin and his theory of evolution.

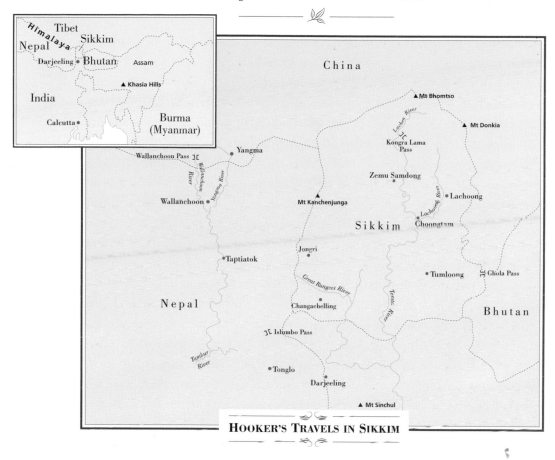

HOOKER'S TRAVELS IN SIKKIM

Hooker arrived back in England on 9 September 1841 and was now determined to make a detailed study of tropical botany so that he could compare it with that of the frozen Antarctic. His friend Hugh Falconer, who was leaving to become Superintendent of the Calcutta Botanic Gardens, and Lord Auckland, the First Lord of the Admiralty, independently suggested Sikkim in the northeast Himalayas, a remote unvisited region, 'ground untrodden by traveller or naturalist'. Hooker applied for and was granted by the Treasury £400 per annum for two years' plant hunting in Sikkim on behalf of Kew.

He left Southampton on 11 November 1847 aboard the steamship *Sidon*, bound for Alexandria. On the voyage he was befriended by Lord Dalhousie, who was travelling out to India as the newly appointed Governor-General. Lord Dalhousie insisted that Hooker accompany him on the frigate sent out to convey him from Suez to Calcutta, and once there he virtually ordered him to stay at his official residence, Government House. The bond that developed between these two highly intelligent but very different men was to prove of great value to Hooker throughout his time in Sikkim, so much so that Hooker named his finest rhododendron after Lady Dalhousie.

After being delayed in Calcutta by the monsoon rains, Hooker began the trek across the fertile plains to the foothills of the Himalayas, arriving at Darjeeling on 16 April 1848. He was immediately befriended by Bryan Hodgson, scholar and foremost Indian zoologist, and Dr Archibald Campbell, the political agent to Sikkim, who mediated between the British Government and the Sikkim Rajah (king). If it was Lord Dalhousie who cleared the official barriers and was to make Hooker's explorations possible, it was Hodgson's practical support throughout the next two years that made them viable.

Hodgson kindly offered Hooker the use of his house as a base while he was in Darjeeling. Looking out from the verandah, Hooker could see a distant sky-line of towering, snow-capped peaks far to the north. In between lay Sikkim, an inhospitable country squeezed between Nepal to the west, Tibet to the north and Bhutan to the east. Hooker wanted to travel to Sikkim's high mountain passes, but to enter this kingdom he required permission from the Rajah. Campbell had spent twelve years trying to foster a better rapport between the British and Sikkimese authorities, but when Hooker arrived relations were still very frosty. Hooker continually applied for a permit, but the Rajah stalled all negotiations. The frustrations continued all summer, but fortunately for Hooker there was much to take his mind off the delays. He completed his *Flora Antarctica* and made several trips to the hills surrounding Darjeeling. In early May Hooker first experienced the thrill of discovering new rhododendrons. On the top of Mt Sinchul, a few miles south-east of Darjeeling, he found the ivory-white-flowered *Rhododendron grande* (*R. argenteum*), which he lavishly described as a 'great … tree forty feet high, with magnificent leaves twelve to fifteen inches long, deep green, wrinkled above and silvery below … I know nothing of the kind that exceeds in beauty the flowering branch of *R. argenteum*, with its wide spreading foliage and glorious mass of flowers.'

He also came across the enchanting *Rhododendron dalhousiae*, 'a slender shrub bearing from three to six white lemon-scented bells, four and a half inches long and as many broad at the end of each branch'. It was growing as an epiphyte on another new find, *Magnolia campbellii*. Magnolias had been introduced into Britain in the late seventeenth century, but the beautiful flowers of this species exceeded all those in cultivation. Hooker described it as 'an immense … sparingly branched tree, leafless in winter, and also during the flowering season, when it puts forth … great rose-purple cup-shaped flowers'.

RHODODENDRON DALHOUSIÆ, Hook. fil.
(in its native locality)

A drawing of *Rhododendron dalhousie*, from *Rhododendrons of the Sikkim–Himalaya* by Joseph Hooker. Hooker thought that this was the most beautiful of his discoveries and named it after Lady Dalhousie.

With its cinnamon-coloured bark and creamy yellow blooms, *Rhododendron falconeri* must have stopped Hooker in his tracks when he first came across it in full flower.

Later the same month, on a journey to Mt Tonglo, Hooker discovered the lovely creamy yellow-flowered *Rhododendron falconeri* in the valley of the Great Rungeet river.

The problems of access into Sikkim continued, and Hooker appealed directly to Lord Dalhousie, who in September demanded that the Rajah give Hooker 'full leave to travel to the Snowy passes and to grant … every assistance'. A rude and flat refusal from the Rajah to the Queen's representative in India provoked the threat of British military intervention, with the result that access was grudgingly granted. In stark contrast, Jung Bahadoor, the Rajah of Nepal, immediately gave Hooker unconditional access to his unexplored eastern border and even provided an escort of six Gurkhas, two officers and a corporal.

For his first expedition into the unknown, Hooker's unusual combination of scholar and adventurer was to prove invaluable. His determination and single-mindedness in the cause of science enabled him to endure great hardship, and the arduous expeditions in Sikkim would test his character and skills to the limits, as the weather, the physical environment and the political situation conspired against him. But through sheer perseverance, good humour and some bullying he succeeded beyond all expectations. A fine illustration of Hooker's professionalism was a telegram he received in 1903 from the Sikkim–Tibet Boundary Commission. It congratulated Hooker on the value and accuracy of the map he had surveyed and drawn some half a century earlier.

Almost a year after leaving England, Hooker finally set out for Sikkim at noon on 27 October 1848. His excitement is clear from a letter to his father: 'I cannot tell you how comfortable I feel at the prospect of realizing the fondest dream I ever harboured as a traveller and botanist.' Following the Tambur river northwards, Hooker proceeded up both its western and eastern forks to the high-altitude Wallanchoon and Yangma passes, 30 and 20 miles to the west of Kanchenjunga respectively. At 28,250ft, Kanchenjunga was then thought to be the world's highest peak.

He was accompanied on the trip by a team of fifty-six (later reduced to fifteen locals), who acted as porters, collectors and guards. Like many of his fellow countrymen at the time, Hooker had a rather condescending attitude towards indigenous people. On the whole he considered them to be uncouth and surly and often commented on their 'slovenly' appearance and lifestyle. Hooker tended to praise the locals only if they were of use to him. He liked the Lepchas, for they were loyal workers, happy to carry loads weighing 80–100 lb for 16 mile stretches. On the other hand, 'I dislike the Bengalees very much; and these are lazy dogs, as all are.' Hooker did, however, have a sensible and self-deprecating approach to dealing with local inhabitants: 'I have always found frankness and kindness good policy with any nation, especially if combined with a reasonable amount of personal vanity, which I abundantly possess, and assumption of superiority and, above all, a liberally flattering opinion of the people openly expressed.' His approach was so successful that the Lepchas still remembered him with great fondness many decades later. In

'Tambur River at the Lower Limit of Pines' – a sketch by Hooker showing the spectacular Himalayan landscape, which he explored for new plant specimens.

return Hooker always noted the many acts of kindness that would be shown to him throughout the difficult years ahead, and took a positive view of the cultural similarities between east and west.

The route to the passes was demanding. There were no proper roads to follow and all progress was made on foot. The rocky, narrow paths wound through tangled shrubs and boulders, brambles, thorny bushes and nettles. The party ascended spurs to 5,000ft or more before plunging into deep valleys, where rushing torrents had to be forded or crossed on frail bridges. In places the valley sides were unstable, and landslides, often up to a mile long, were a constant danger to the expedition.

As winter approached, the conditions deteriorated and the temperature fell steadily as the party climbed higher. The weather could swiftly turn. A mountainside that was bathed in sunshine could be blanketed by fog or lashed by rain, sleet or snow in under an hour. Keeping dry was a continual problem: there was no shelter in the sodden forests, which constantly dripped condensed fog, dry firewood was hard to find and the oxygen-deficient air meant fires burned poorly. To add to these discomforts, Hooker suffered from altitude sickness and at times found taking only a few footsteps extremely tiring. Splitting headaches and dizziness hampered his ability to record details and sunshine reflecting off newly fallen snow caused sore eyes. Nevertheless, he followed a taxing routine – he walked from 10 a.m. till 4 or 6 p.m. every day, collecting plants along the way. In the evenings he wrote up his journal and notes, plotted maps and catalogued the day's finds by lantern light.

After the small luxury of his evening cigar, Hooker retired to a tent of blankets spread over cross-pieces of wood and a ridge pole. But even this small haven presented him with further problems. He frequently had to build a wall of turf or stone outside the entrance to his tent to guard against the bitterly cold wind which blew down from the glaciers. Insects were a continual nuisance as 'large and small moths, cockchafers, glow-worms, and cockroaches made my tent a Noah's ark by night, when the candle was burning; together with winged ants, May-flies, flying earwigs, and many beetles, while a very large species of *Tipula* (daddy-long legs) swept its long legs across my face as I wrote my journal, or plotted off my map'. The Tibetan dogs were let loose at night and stole into the camp looking for food, and the yaks used by the locals as pack animals would push their inquisitive heads into the tents throughout the hours of darkness. Hooker became tired of being woken up by snorts of moist air and slept with a heavy tripod by his side 'to poke at intruders'.

He recorded with awe the immense peaks, thick forest, broad flat valleys, brilliant starlit nights and majestic sunrises. In the lower Myong valley (8,000ft), the hill crests were clothed in cheer pine (*Pinus longifolia*), clumps of oak and other trees, bamboos and bracken. The valley sides were cut by little ravines, which were carpeted with lush tropical vegetation, and pebbly streams of crystal clear water which flowed down from the heights. As they travelled further north, beyond Taptiatok (also 8,000ft), the scene became more rugged: 'as grand as any pictured by Salvator Rosa; a river roaring in sheets of foam, sombre woods, crags of gneiss, and tier upon tier of lofty mountains flanked and crested with groves of black firs, terminating in snow-sprinkled rocky peaks'.

Hooker arrived at Wallanchoon, 10,385ft, on 23 November. The village contained about 100 scarlet-painted wooden houses, measuring up to 43ft high and 82ft long and accommodating several 'good-natured, intolerably dirty' Tibetan families. Each family lived in a couple of rooms with an open fire in the largest. The villagers were 'Mongolian in countenance' and the buildings were surrounded by long poles and vertical flags and rhododendron bushes. On 26 November he reached the western Wallanchoon Pass (16,748ft), where the air temperature was a numbing 18°F. Hooker then followed the eastern branch to the Yangma Pass until snow at 15,168ft prevented any further progress. The scenery of the Yangma valley was 'wild and very grand, [the] path lying through a narrow gorge, choked with pine trees, down the river roared in a furious torrent: while the mountains on each side were crested with castellated masses of rock, and sprinkled with snow. The path was very bad, often up ladders, and along planks lashed to the faces of precipices, and overhanging the torrent, which it crossed several times by plank bridges.'

This valley and Mt Nango provided Hooker with a rich hunting ground for new species. His description of the graceful Sikkim larch does not do the plant justice: 'It is a small tree, twenty to forty feet high, perfectly similar … to a European larch, but with larger cones … its leaves – now red – were falling and covering the rocky ground.' However, he was more enthusiastic about *Rhododendron hodgsonii*, which he first saw in a glorious setting: 'The ground was covered with silvery flakes of birch bark, and that of *Rhododendron hodgsonii*, which was as delicate as tissue paper, and a pale flesh-colour … I was astonished at the beauty of its foliage, which was of a beautiful bright green, with leaves sixteen inches long.' Two more rhododendrons were added to his tally: the striking *R. thompsonii* and the graceful *R. campylocarpum*. Unlike the

Rhododendron thomsonii, shown here in a drawing by Walter Fitch,
from *Rhododendrons of the Sikkim-Himalaya* by Hooker.

poisonous *R. cinnabarinum* (see page 93), the foliage of both species was eaten by sheep and the natives sometimes ate the flowers of the former.

Hooker now hoped to cross the Kanglanamo Pass, but the season was late and the pass was blocked with snow. Camping became a perilous occupation, with heavy snowfalls threatening to bury him alive, and he had to take the precaution of stretching a tripod stand over his head to leave a breathing hole should the roof cave in. He was forced to travel south to the Islumbo Pass and on reaching Lingcham (4,860ft), where the weather was 'gloomy, cold, and damp, with much rain and fog', he received the cheering news that he was to meet Archibald Campbell at Bhomsong, where a durbar (summit) was to be held between the British and the Rajah. Heading east, Hooker left Lingcham on 20 December accompanied by his friend the Kajee (head man) who was 'constantly followed by a lad, carrying a bamboo of Murwa beer slung round his neck, with which he [the Kajee] kept himself always groggy. His dress was thoroughly Lepcha, and highly picturesque, consisting of a very broad-brimmed round-crowned bamboo-platted hat, scarlet jacket, and blue-stripped cloth shirt, bare feet, long knife, bow and quiver, rings and earrings, and a long pigtail.'

The Kajee confided to Hooker that it was his spectacles that commanded universal respect in Sikkim because they made him look wise. He asked Hooker to send him a pair, which he did and which the Kajee wore with great pride on state occasions. On the way to Bhomsong Hooker bought a black puppy (a Tibetan mastiff, a common Sikkim hunting dog cross), which he named Kinchin. This faithful hound became his constant companion in the lonely months in Sikkim.

The intention of the durbar was to forge better relations between the two countries, but unfortunately the Rajah was dominated by his Dewan (prime minister). The Dewan was 'neither financier nor politician, but a mere plunderer of Sikkim' who wanted to control Sikkimese trade. He saw the British as a threat to his plans and succeeded in scuppering the negotiations. Following the breakdown of talks, Hooker reluctantly 'bade adieu' to Campbell on 2 January 1849 and completed the last leg of his journey. He backtracked west to Jongri, a deserted Yak post at 13,000ft. Visiting the Buddhist temples at Changachelling, which were being redecorated, he was amused to see among the figures an Englishman 'depicted in a flowered silk coat instead of a tartan shooting jacket, my shoes were turned up at the toes, and I had on spectacles'. On 19 January Hooker arrived back in Darjeeling, where he was surprised to be woken on the first morning by the sound of wildlife. He was much more accustomed to waking up to the sounds of the lamas chanting and playing drums, gongs and trumpets. The 'wild music', as he called it, deeply impressed him but he found that it also depressed him, emphasizing as it did the foreignness of his situation.

Hooker's collection was so great that it took him six weeks to arrange, catalogue and pack the '80 coolie-loads' of specimens that were sent back to his father at Kew. Spurred on by his success, he planned a second expedition, this time to the passes east of Kanchenjunga. In the meantime he over-wintered in Darjeeling, taking advantage of 'Hodgson's society and library [and] Campbell's zealous interest'. The second expedition left Darjeeling on 3 May 1849, following the Teesta river up steep valleys to its headwaters and then ascending both forks: the western fork was called the Lachen and led to the Kongra Lama pass; the eastern, known as the Lachoong, led to Donkia Pass. The journey was to yield spectacular results but nearly ended in catastrophe.

Hooker's progress was constantly impeded by the Dewan, who did not want this interfering Englishman back in his country. His agents sabotaged the party's food supplies whenever possible, damaged paths and bridges ahead, and forbade native villagers to sell Hooker provisions. This did not, however, stop the locals showing him considerable and much appreciated kindness, and at great risk to themselves, they often left gifts of food for him and his team. Throughout his tribulations the young Englishman managed to retain his sense of humour. On one occasion when they were out of food a flock of wild sheep suddenly appeared. Hooker wryly noted: 'I could not but bitterly regret the want of a gun, but consoled myself by reflecting that the instruments were still more urgently required to enable me to survey this extremely interesting valley.' Another time he was reduced to eating coarse boiled rice and chilli vinegar for eight consecutive days. However, he refused to be intimidated and went on the counter-offensive, using a combination of 'bullying, douce violence, persuasion, prescribing of pill, prayers and charms' to aid his progress. Nonetheless, the Dewan's constant interfering meant that an expected thirty-day trek from Darjeeling to the Kongra Lama pass took eighty-three days.

Travelling at this time of the year was especially hazardous. The spring melt meant an increased danger of landslip and avalanche – at night the roar of falling rocks and ice carried on for hours on end, and one morning Hooker awoke to find that a large boulder had narrowly missed his tent. The weather and the insects also caused considerable discomfort. The temperature was stiflingly hot and the rains, which had broken on 10 May, brought scant relief and left the party permanently damp. Hooker was plagued by swarms of gnats, microscopic midges and peepsas (small biting flies). However, it was leeches that caused the greatest suffering. He complained that his legs were daily clotted with blood and he 'repeatedly took upwards of a hundred from my legs, the small ones used to collect in clusters on the instep'. The continual problems eventually wore him down and depressed his normally optimistic and positive spirit. As he rather forlornly wrote:

> The isolation of my position, the hostility of the Dewan, and consequent uncertainty of the success of a journey that absorbed all my thoughts, the prevalence of fevers in the valleys I was traversing, and the many difficulties that beset my path, all crowded on the imagination when fevered by exertion and depressed by gloomy weather, and my spirits involuntarily sank as I counted the many miles and months intervening between me and my home.

At the village of Choongtam, where the Teesta river divides, Hooker was bemused by the welcome of the locals, many of whom had never seen a westerner before. It was not until he was back in Britain that he discovered that the lolling of tongues and the pulling of ears was a traditional Tibetan salute. While waiting for food supplies, a problem that caused the party to have to be reduced from forty-two to fifteen, Hooker collected ten types of rhododendron, including two more new species: the stunning *R. griffithianum (aucklandii)* – the most important in terms of later hybridizing – and the richly-scented *R. edgeworthii*, which 'in the delicacy and beauty of their flowers … perhaps, excel any others'. From here he first climbed the western stream, the Lachen. The unwilling guide of the party, whom he suspected of taking the most awkward route, led them to a hazardous river crossing. Undaunted, Hooker:

> took off my shoes, and walked steadily over: the tremor of the planks was like that felt when standing on the paddle-box of a steamer, and I was jerked up and down, as my weight pressed them into the boiling flood, which shrouded me with spray. I looked neither to the right nor to the left, lest the motion of the swift waters should turn my head, but kept my eye on the white jets d'eau springing up between the woodwork, and felt thankful when fairly on the opposite bank: my loaded coolies followed, crossing one by one without fear or hesitation. The bridge was swept into the Lachen very shortly afterwards.

At Zemu Samdong three productive days were spent collecting 'many new and beautiful plants', including the handsome *Rhododendron niveum* and the exotic *R. cinnabarinum*. The latter, Hooker observed, was toxic. When burned as fuel, the smoke caused inflammation of the face and eyes, the nectar made poisonous honey and 'goats and kids … died foaming at the mouth and grinding their teeth … from their eating the leaves'. As well as the rhododendrons, Hooker found two of his most charming herbaceous species. Growing by the streams were *Primula sikkimensis*, the 'magnificent yellow cowslip [which] gilded the marshes', and swathes of the mauve-flowered *P. capitata*.

His high spirits returned on the afternoon of 24 July, when he reached the cairns on the Kongra Lama Pass at an altitude of 15,745ft, marking the border with Tibet. Although bitterly cold, he collected forty kinds of plant and gazed at the great mountains on either side. On the night-time journey back to camp Hooker was:

> at liberty uninterruptedly to reflect on the events of a day, on which I had attained the object of so many year's ambition. Now that all obstacles were surmounted, and I was returning laden with materials for extending the knowledge of a science which had formed the pursuit of my life, will it be wondered at that I felt proud, not less for my own sake, than for that of the many friends, both in India and at home, who were interested in my success?

Among the party was the unenthusiastic Singtam Soubah (the head man of Singtam), sent by the Dewan to 'aid' Hooker. During the short period that the Singtam remained with him, Hooker was able to build up a friendship (which was later betrayed) by curing the man's upset stomach. In return he was treated to the enjoyable Tartar hospitality, which included serving salted and buttered tea and popcorn.

Hooker found so many new rhododendrons that it is easy to overlook some, but the red bell-like flowers of *Rhododendron cinnabarinum* ssp. *cinnabarinum* are a striking sight.

Returning to Choongtam on 5 August and refreshed after ten days' rest, Hooker set off up the eastern stream, the Lachoong. It was here that his dog, Kinchin, tragically fell from a cane bridge into the roaring torrent below. Hooker was deeply distressed and 'for many days I missed him by my side, and by my feet in camp'. Continuing northwards, he reached the Donkia Pass at 18,466ft on 9 September. Here he noted that: 'Never, in the course of all my wanderings, had my eye rested on a scene so dreary and inhospitable.' Hooker climbed Mt Donkia in order to get a better view of the surrounding mountains, and unknowingly broke a second world record when he ascended to 19,300ft. As if nature was celebrating his achievement, he was treated to a display of the 'spectre of Brocken' – a rare atmospheric phenomenon by which his shadow was projected on to a mass of thin mist above him and his head was surrounded 'with a brilliant circular glory or rainbow'.

On 5 October Hooker met up with Archibald Campbell at Choongtam. The purpose of Campbell's presence in Sikkim was once again to try to foster better relations and to investigate Hooker's ill-treatment. It also gave him the perfect opportunity to become 'acquainted with the country'. Together the men retraced the trek to the Kongra Lama Pass. This time, rather than stopping at the border, Hooker became an 'illegal immigrant' as he goaded his pony into Tibet, in spite of the border patrol. After some quick talking the remainder of the party was allowed into Tibet, but was escorted by the border patrol, the commander mounted on a yak. Hooker and Campbell climbed Mt Bhomtso, reaching 18,590ft in terrible weather. A ferocious dry north-west wind blew, 'peeling the skin from our faces, loading the hair with grains of sand'.

The party returned to Sikkim via the Donkia Pass, and on the way down to Choongtam Hooker discovered that he had lost one of his thermometers. It had last been seen at the hot springs below the Kinchinjhow glacier, and Cheytoong, the lad responsible for looking after it, was so distraught that despite attempts to stop him he turned back alone and went to look for it. Three days later Hooker was relieved to see him return, carrying with him the precious thermometer. It transpired that rather than give up his search and descend from the mountain when darkness fell, Cheytoong had spent the whole of the freezing October night lying in the hot waters of the springs.

Having successfully completed the circuit that he had been unable to do on his first trip, Hooker now decided to visit the Chola and Yakla Passes in eastern Sikkim with Campbell. The route passed through the capital, Tumloong, where Campbell expected to meet the Rajah on his return. The path up to the

Chola Pass at 14,925ft produced a bounty of rhododendron seed – Hooker collected twenty-four species: *R. anthopogon*, *R. arboreum*, *R. camelliaeflorum*, *R. campanulatum* ssp. *aeruginosum* (syn. *R. aeruginosum*), *R. campbelliae*, *R. campylocarpum*, *R. ciliatum*, *R. cinnabarinum*, *R. dalhousiae*, *R. edgeworthii*, *R. falconeri*, *R. fulgens*, *R. glaucum*, *R. grande*, *R. hodgsonii*, *R. lanatum*, *R. barbatum* (syn. *R. lancifolium*), *R. lepidotum*, (syn. *R. eleagnioides*, *R. obovatum* and *R. salignum*), *R. niveum*, *R. setosum*, *R. thompsonii*, *R. vaccinioides*, *R. virgatum* and *R. wightianum*. In 'clefts at 10,000 feet' he also discovered *Meconopsis villosa*, 'a beautiful yellow poppy-like plant'.

At the Chola Pass a Tibetan border patrol turned the party back and on the return to the village of Chumanako they were surprised to meet the Singtam Soubah. That night an incident that changed the map of India occurred. When Campbell and Hooker entered the hut where they planned to spend the night, a large group of ugly-looking Sikkim Bhutanese crowded in after them. Campbell decided it would be best if they camped out instead, and left the hut in order to organize the pitching of the tents. Startled by Campbell's cries of 'Hooker! Hooker! the savages are murdering me!' Hooker rushed out to see Campbell fighting a large crowd of men. Hooker tried in vain to help his friend but was restrained by the mob and forced back into the hut while the crowd beat and tortured Campbell. Campbell believed he was about to be murdered, as he was first trampled on and then had his head violently pressed down on to his chest. His arms were tied behind his back and a bamboo stick was used to tighten the cord. The treacherous Singtam Soubah, shaking with nerves, informed Hooker that Campbell was a political prisoner by order of the Rajah, who, unhappy with Campbell's conduct over the past twelve years, wanted to hold him hostage until the British conceded to his demands. In reality the Rajah knew nothing of the kidnap, and the Singtam had betrayed his friend and Campbell to the Dewan. The Dewan's somewhat flawed reasoning was that harming a British official would bring about a lasting breach in relations, resulting in no more foreign interference. This would give the Dewan time to increase his monopoly on Sikkimese trade, which he would defend with his own army of renegades.

Hooker was offered his freedom, but he was determined to stay with his friend. Escorted back to Tumloong, Campbell was kept under heavy guard while Hooker 'kept as near as I was allowed, quietly gathering rhododendron seed by the way'. While Campbell was treated with open hostility, Hooker was constantly reassured that he was not in danger. He was offered a pony to ride on, but when he saw the demeaning way Campbell was forced to walk at the tail of a mule he refused. From 10 November until 7 December they were imprisoned. At first, they were kept separate, but Hooker persuaded his captors to let him move into the small bamboo and wattle hut in which Campbell was being kept and where they slept in a tiny back room that had been Campbell's cage for the first few days of his incarceration.

The capital of Sikkim, Tumloong, with the Rajah's residence in the background, and the hut where Hooker and Campbell were held captive.

The plight of Hooker and Campbell was unknown to the Darjeeling authorities for some time, because the letter sent by the Dewan informing them of Campbell's arrest and his demands was written in Tibetan. Ironically, his secretary, not fully understanding it, placed the letter on Campbell's desk for his attention upon his return! Eventually Hooker got a letter to Lord Dalhousie, who immediately responded by ordering an English regiment and three guns to the border. This action shocked the Dewan and brought him to his senses. Realizing for the first time the full significance of holding the two men hostage and the possible repercussions, he quickly released them. The Dewan marched the men to Darjeeling under heavy guard. Incredibly he also took eighty loads of merchandise, still expecting to trade at Darjeeling on friendly terms. After a nerve-racking last few days, during which they expected to be murdered (to prevent their giving evidence against their captors), the two weary travellers arrived back to celebrations on Christmas Day 1849.

The British authorities were incensed at the ill-treatment of two important subjects and were determined that there should be some form of retribution. Hooker noted that in some previous disagreements with Indian rulers the British Government had not responded to provocation. This time, however, the threat of invasion was real. Hooker was asked to use his knowledge of the terrain to draw up a plan of attack, while the Rajah was ordered to come to Darjeeling, bringing with him the main offenders. The Rajah did not appear, and, although the invasion was not carried out because of poor military decision-making hundreds of miles away, the whole of southern Sikkim was annexed into India, thus adding another small pink corner to the map of the Empire. The annexed area, the only fertile land in Sikkim, proved to be perfect for the future cultivation of tea and quinine.

Hooker now spent three peaceful months recovering in Darjeeling. He completed his map and arranged his botanical collection (which on this trip amounted to some 100 man-loads) before making his last journey to the Khasia Hills in Assam (1 May 1850 to 28 January 1851). This area had been extensively botanized, but Hooker collected seven man-loads of the very beautiful and fashionable blue orchid *Vanda caerulea*. At Myrung he met Thomas Lobb (see Chapter 6).

After three and a half exhausting years in India, exploring one of the most inaccessible terrains in the world, Hooker arrived home to a rapturous welcome on 26 March 1851. Unlike many other plant hunters, he had no problems adjusting back into life in Britain and on 15 July 1851 he married his patient fiancée, Frances. Stung by the refusal of a government grant to finish his work, and already £800 down on his trip to Sikkim, he acidly remarked that had he been a commercial plant hunter he could have made '£1500 by Rhododendron seed and seedlings alone'. After much persuasion, a three-year grant of £400 per annum was forthcoming, and this enabled him to complete his *Flora of New Zealand* (1853) and write his *Himalayan Journals* (1854). To quote Mea Allen, the *Journals* 'with Wallace's *Malaya Archipelago* and Charles Darwin's *Voyage of the Beagle* form a trilogy of the golden age of travel in pursuit of science'. Hooker now renewed his friendship and correspondence with Charles Darwin, becoming his respected critic and confidant. The admiration was mutual, and during the sixteen years between their first correspondence and the publication of *Origin of Species*, Hooker never betrayed the secret of Darwin's theory of evolution.

In 1855 Hooker's overworked father, Sir William, finally persuaded him to become his assistant director at Kew. Under Sir William's guidance and vision Kew was going through a period of enormous development: it increased in size from 15 to over 250 acres within five years. In 1848 both the Palm House, designed by Decimus Burton, and the Museum of Economic Botany were opened; the library was inaugurated, and William's private collection of dried plant specimens became the nucleus for what is now the world's largest herbarium. Kew was not solely a scientific institution, however. The beautifully landscaped gardens and extensive collection of glasshouses were very popular with the public, and between 1841 and 1850 visitor numbers increased almost twenty-fold, from 9,174 to 179,627. Interest continued to grow, and at the turn of the century nearly 3 million visitors a year were passing through the gates.

As assistant director, Joseph was responsible for much of the day-to-day running of Kew, including the organization of the enormous herbarium collections, which were coming in either as gifts from private individuals or from the plant hunters. Among these was his own collection from India and Sikkim, which he combined with that of Dr Thomas Thompson who had botanized the north-west Himalayas and Tibet. This joint collection contained nearly 7,000 species and included the Sikkim rhododendrons. The first chance the world at large had to admire these sensational plants was the publication of *The Rhododendrons of Sikkim–Himalaya* (1849–1851), edited by Sir William Hooker from notes made in the field by Joseph, which contained beautiful and detailed coloured plates painted by Walter Hood Fitch.

In 1851 these evergreen broad-leaved flowering trees had come back to Britain as minute seeds. Now a privileged few across Britain were given seed and/or seedlings by Kew. All hoped to be the first to flower the new rhododendrons and spent much time and money nurturing these precious plants, reporting their successes and failures back to Kew. In 1857 Bagshot Nursery, near Sunningdale, managed to get the brilliant red *R. thompsonii* to bloom by grafting scions on to established glasshouse rootstock. Bagshot was the only nursery which offered the full range of Robert Fortune's China collection (see Chapter 5) and a wide range of David Douglas's finds (see Chapter 3). Quite how Bagshot came to obtain such rarities is unclear, but the nursery was instrumental in popularizing many new introductions.

With such enormous public interest, rhododendrons were now poised to make their impact on nineteenth-century garden fashion. Paradoxically, the catalyst for this craze was the tenderness of the new species. From reports reaching Kew it soon became apparent that they thrived outdoors only in the mildest and wettest parts of Britain, particularly Cornwall and the west coast of Scotland. Once this was recognized, nurseries such as Bagshot and its neighbour, Waterer's, lost no time in crossing the new Himalayan species with their hardy cousins. The four most important hybrid parents were *R. campylocarpum*, *R. ciliatum*, *R. thompsonii* and *R. griffithianum*, the last being the most significant. The early results were spectacular: a myriad hardy hybrids with glorious foliage and vibrant flowers, appealing strongly to the Victorian desire for novelty, bold colour and a clear display of art in the garden. Building a collection of rhododendrons became an obsession for many wealthy landed gardeners, and the designed landscape around many British country estates changed dramatically as a result.

Many of the species already grown in the 'American garden' required moist, acid soil – the perfect conditions for the new rhododendrons, which were soon added in large numbers. Such a mixture of plants, if no longer a purely geographic collection, had its benefits. The spring-flowering display was spectacular, as was the contrast in foliage form, texture and colour between the evergreen shrubby rhododendrons and the conifers. Rhododendromania took such a grip that by the end of the 1860s rhododendrons had almost entirely replaced the American flora (although a few species such as *Kalmia latifolia* and *Gaultheria procumbens* were retained), and eleven years later Shirley Hibberd wrote that 'The money spent on rhododendrons during twenty years in this country would nearly suffice to pay off the National Debt.'

Although adding rhododendrons to existing garden features was popular, the 'purist' restricted rhododendrons to a new type of theme garden – the rhododendron garden. There was debate, however, on how best to display a rhododendron collection. One method was to plant specimens to be admired alone, as, for example, at Menabilly in Cornwall (the house where Daphne du Maurier set *Rebecca*), where an area of 2 acres called Hooker's Grove was studded with Himalayan rhododendrons. An alternative was to mass-plant rhododendrons in beds or shrubberies. At Oaklands in Surrey, for example, thousands of rhododendrons were planted in large circular beds surrounded by grass. The rhododendrons were gathered together in clumps of twenty or more varieties and were carefully selected to maximize the flowering season.

The restricted flowering season was one of the main problems of the rhododendron garden, and some writers suggested introducing other exotics, such as conifers and lilies, which 'lift their noble heads out of the rich green beds, and fill the air with fragrance', to lengthen the season of interest.

As the change towards naturalistic planting took hold in the late nineteenth century, rhododendrons were the source of inspiration for another innovation. They were blended with native and naturalized species and other exotics in semi-natural features such as shrubberies, tree-belts and woodland plantings. The plants were carefully selected and arranged so that they could be individually admired, but the overall effect was greater than the sum of the parts. These features were scattered throughout the landscape yet carefully integrated into the plan as a whole, to create a display that was constantly changing, but always presented a perfectly balanced picture. Once again Cornwall offers one of the best examples. At Lamorran, where the magnificent *R. griffithianum* first flowered outdoors, the grass slopes that flowed down the sheltered valley garden to the sea were carefully planted with thick clumps of rhododendrons and other exotics. The effect was so impressive that in 1877 the garden was described as 'a veritable store house of rich and rare, so skilfully cultivated and arranged in such good taste as to be always fresh and always attractive'.

The 1880s saw yet another new fashion – the creation of replica foreign scenery. This also had its origins in the idea of natural or wild gardening, and by far the most popular 'imitation' was the dramatic rhododendron forest-cum-Himalayan valley inspired by Hooker's *Journals*. A spectacular example was the aptly named Cragside in Northumbria, where by the 1890s several hundred thousand rhododendrons had been planted over a rocky valley side. These formed 'impenetrable thickets … blooming so profusely as to light up the whole hillside with their varied colours.'

Although Hooker had the woodland around his home, The Camp near Sunningdale, planted extensively with 'his' rhododendrons, he seems to have taken little interest in the evolution of the new garden fashions. By now he was successful, happily ensconced at Kew, and had begun working with George Bentham on the *Genera Plantarum*. This *magnum opus* took twenty-six years to complete (the last part was published in 1883), and it became the most outstanding botanical work of the nineteenth century. In 1,681,500 words on 3,363 pages (not including the 200-page index) the authors described all the known members of 200 plant families, all of which had been individually examined. At last Banks's dream of having a record of every plant known to exist in the colonies was fulfilled (although additions would be required as more new species arrived).

On 1 November 1865 Joseph succeeded Sir William as Director of Kew. Following in his father's footsteps was never going to be easy, but he proved himself an equally capable Director. He was particularly interested in the fields of taxonomy, economic botany and education. In 1876 he established the Jodrell Laboratory for research into plant physiology and anatomy, and by the turn of the century 700 Kew-trained botanists and gardeners were working around the world, often in posts appointed by Kew. Despite his hectic schedule, Hooker found time to indulge his passion for travel. In the autumn of 1860 he visited the Lebanon to study the cedars; in 1871 he went to Morocco, climbed the Atlas Mountains and found a species of annual toadflax (*Linaria maroccana*). A sprightly sixty-year-old, he made his last expedition in 1877, visiting Colorado, where he climbed to nearly 13,000ft in the Rocky Mountains in order to compare the alpine flora with that of Sikkim. The same year, having declined it twice before, he finally accepted a knighthood.

Eight years later Hooker, and his doctors, felt that the time had come for him to pass control of Kew on to his son-in-law, W. T. Thiselton-Dyer. This did not mean, however, that he was inactive during his retirement. Over the course of the next twenty-six years he wrote at a prolific rate, completing *The Flora of Ceylon* and *Flora Indica* among others, and he continued to classify plants. He died peacefully at the age of ninety-four while working on the balsams, in his words, those most 'deceitful above all plants'.

Botany owes Joseph Hooker an enormous debt, not only for his lifelong study of plants but also for reaffirming Kew's position as the world's foremost centre for botanical study. His impact was literally on a global scale – the British Empire flourished as a result of his concerted efforts to grow economically important crops in the colonies, and spring gardens around the world are lit up by the descendants of the wonderful Himalayan rhododendrons he discovered.

Joseph Hooker's Plant Introductions

The date beside each plant name is the date of its introduction into Britain.

Primula capitata (1849)

Primula = from (Lat.) *primus*, first, i.e. early flowering

capitata = (Lat.) growing in a dense head, i.e. the flowers

Bold drumsticks of deep purple flowers are carried on 4–18in (10–45cm) long powdery stems from July to September. A remarkable late-flowering primula with rough, powdery leaves. George Forrest introduced the subspecies *sphaerocephala* (*c.* 1910) from Yunnan, with powderless leaves.

Ranges from east Nepal to Sikkim, Bhutan, southeast Tibet, north-west Burma and Yunnan in China, growing in a variety of alpine habitats above the tree-line at considerable elevations from 10,000–18,000ft (3,000–5,500m).

Primula sikkimensis (1849)

Primula = from (Lat.) *primus*, first, i.e. early flowering

sikkimensis = (Lat.) of Sikkim

Deliciously fragrant, soft yellow, nodding flowers appear from May to July on 6–35in (15–90cm) stems above neat rosettes of grey-green leaves. The magnificent Himalayan cowslip is a variable species and heads the *Sikkimensis* group of primulas, which includes such choice plants as *P. alpicola*, *P. florindae* and *P. waltonii*.

Hooker is best known for his rhododendron discoveries, but he also found several beautiful herbaceous species, including *Primula capitata*.

Occurs from west central Nepal through Sikkim, Bhutan and north Burma and east to Yunnan and south Sichuan in China, carpeting large areas of streamside and bog in glacial valleys at 9,000–16,000ft (2,900–5,200m).

Rhododendron cinnabarinum (1849)

Rhododendron = (Gk) *rhodo*, rose; *dendron*, tree
cinnabarinum = (Lat.) cinnabar-red, vermilion

Exquisite cinnabar-red, pendant, tubular bells are borne in May to June among vivid turquoise-green, aromatic foliage. This beautiful and variable open bush has varieties of many shades, from apricot-orange to crimson or rich plum-purple. All forms are poisonous, and the nectar even contaminates honey. Hooker introduced two forms, *R. c.* subsp. *cinnabarinum* and *R. c.* Roylei Group (plum-crimson).

Occurs as one of three subspecies from Nepal to Bhutan, south-east Tibet and Upper Burma, growing in mixed woodland, conifer and rhododendron forests or rocky hillsides at 6,800–13,000ft (2,100–4,000m), reaching 10–20ft (3–6m) tall.

Rhododendron falconeri (1850)

Rhododendron = (Gk) *rhodo*, rose; *dendron*, tree
falconeri = after Hugh Falconer, Scottish botanist

Magnificent 8–14in (20–35cm) long, dark green, leathery leaves are heavily veined and felted beneath with rich cinnamon. This superb large shrub or small tree has attractive reddish bark and bears great candelabra of creamy-yellow, bell-shaped flowers in April to May.

Ranges from east Nepal to Bhutan and west Arunachal Pradesh, in mixed forest at 8,860–11,000ft (2,700–3,400m), often in colonies and with *R. hodgsonii*, reaching 39ft (12m) tall.

Rhododendron griffithianum (1850)

Rhododendron = (Gk) *rhodo*, rose; *dendron*, tree
griffithianum = after William Griffith, botanist

Glorious, fragrant white, lily-like flowers up to 6in (15cm) across in trusses of 3–6 appear in May. A very fine large shrub with ample 12in (30cm) long leaves and smooth, peeling, multi-coloured bark. It is a key species in rhododendron breeding.

From east Nepal, Sikkim and Bhutan to south-east Tibet and east Arunachal Pradesh, growing in moist forests of oak, magnolia and rhododendron at 5,700–9,000ft (1,800–2,900m), reaching up to 20ft (6m) tall.

Rhododendron hodgsonii (1850)

Rhododendron = (Gk) *rhodo*, rose; *dendron*, tree
hodgsonii = after Bryan Houghton Hodgson, naturalist

Sumptuous, waxy, bell-like flowers of pink or reddish-purple in dense trusses above 10in (24cm) leathery leaves in March to May. This handsome large shrub has lovely flaking bark, and the wood is popular for carving.

Ranges from east Nepal, Sikkim, Bhutan and west Arunachal Pradesh, growing in pine and silver fir forests with bamboo and rhododendrons (*R. arboreum*, *R. barbatum* and *R. falconeri*) at 9,000–14,000ft (2,900–4,300m), reaching up to 20–30ft (6–9m) tall.

Rhododendron thomsonii (1850)

Rhododendron = (Gk) *rhodo*, rose; *dendron*, tree
thomsonii = after Thomas Thomson, Superintendent of Calcutta Botanical Gardens

Intense blood-red flowers appear in April to May, followed by attractive green and glaucous fruiting clusters. This beautiful shrub has attractive, smooth peeling bark in shades from cinnamon to creamy-brown or maroon, and rounded green leaves, which are glaucous when young. The vivid flower colour has been incorporated into some fine hybrids e.g. 'Cornish Cross', 'Aurora' and 'Shilsonii'.

Ranging from east Nepal and Sikkim to Bhutan, southern Tibet and Arunachal Pradesh, growing in dense rhododendron and conifer forest, bogs and streamsides at 8,000–14,000ft (2,500–4,300m), reaching 3–22ft (1–7m) tall.

5

FORTUNE FAVOURS THE BRAVE

Robert Fortune

(1812–80)

I N 1841 BRITAIN WAS BEGINNING TO EMERGE AS THE WORLD'S MOST powerful nation. However, there was a price to pay for the Empire's ascendancy. Three-quarters of a century of industrialization had generated a rash of social ills such as slum housing, terrible public health and the appalling use of child labour. Abroad, colonial expansion continued, and as Sir William Hooker travelled south from Glasgow to the Royal Botanical Gardens at Kew, on the other side of the world British naval warships were helping to win the Opium Wars against China.

Ever since the first Western traders, the Portuguese, had reached China by boat in 1516, the mysterious Orient had held Europe in awe. Tales brought

---

Left: Fortune was deeply impressed by much of China's landscape and wrote enthusiastically about its beauty.

Above: A formal portrait of Robert Fortune, who spent almost twenty years exploring the Orient.

back by explorers and adventurers increased China's mystique, but until the nineteenth century her enormous size and strength held at bay the appetites of countries such as Portugal, Spain and Holland. Now, to a technologically advanced Britain, China was too rich a picking to ignore any longer.

Trade links with China had been established by the 1830s, but the Chinese understandably regarded the profiteering West with suspicion and mistrust, and maintained a tight control over their contact with merchants. This did not suit the British, who precipitated the Opium Wars of 1839–42 in order to force a trade opening through a commercial treaty. The Treaty of Nanking, signed in 1842, had two major implications. First, the small, barren island of Hong Kong was ceded to the British, and four mainland treaty ports – Shanghai, Ningpo (now Ningbo), Foochow (now Fuzhou) and Amoy (now Xiamen) – were opened to the west. This enabled much 'improved' trade with China. Second, China was forced to buy Indian opium in exchange for luxury goods such as porcelain and silk, which were then imported into Britain. This opium trade, much disliked by the Chinese authorities, meant that the East India Company (which had been banned by the British Government from selling the drug to the Indians in the 1830s) could save its Bengalese opium industry.

The treaty had a further important implication. Chinese plants such as *Hydrangea macrophylla*, the tree paeony (*Paeonia suffruticosa*) and chrysanthemums had arrived in Britain in the eighteenth and early nineteenth centuries, causing great excitement. Banks had sent William Kerr to China in 1804 and he had returned with many plants, including *Kerria japonica*, the tiger lily (*Lilium lancifolium*) and the Chinese fir (*Cunninghamia lanceolata*). It was a widely held belief that many more treasures were waiting to be discovered, and now, for the first time, there was an opportunity to carry out peaceful plant hunting in the unexplored northern areas of China. The first person to recognize this was John Reeves (1774–1856), a retired tea inspector, who while working in Canton had sent back plants and seeds to the Horticultural Society (he is credited with the introduction of the beautiful *Wisteria sinensis* in 1816). As the leading light on the Society's influential Chinese Committee, he cleared the way for a plant hunter to travel to China and collect on the Society's behalf.

The man chosen was the thirty-year-old Robert Fortune, another in the long line of successful Scottish plant hunters, who was to explore the Orient for over nineteen years. Little is known of Fortune's early life, and although he kept a detailed journal and wrote frequent letters home, these have been lost. All we are left with to make an assessment of his character are the four books describing his travels, and the articles he wrote for the *Gardener's Chronicle* and the *Journal of the Royal Horticultural Society*. Fortune's style of writing is curiously impersonal, with a directness that, although making for exciting reading, makes it even harder to judge the man, but some conclusions can be drawn. That he brought back so many beautiful plants which were capable of being grown in the garden demonstrates that the Society chose a skilled botanist and horticulturist to do their work. That he learned some Chinese, a difficult language, reveals a quick mind. His extensive travels into areas where he was not permitted to go, using a disguise, shows that he was ingenious, inventive, resourceful and determined. And his reaction when attacked by pirates and facing almost certain death if caught tells us that he was not only courageous but also level-headed and calm in a crisis, as well as a good shot!

In obtaining plants from Chinese nurserymen Fortune demonstrated another defining character trait: he won their trust with his honesty and showed almost superhuman levels of tact and patience in his dealings with them. Although he was frequently robbed (perhaps because he naïvely believed everyone shared his levels of integrity), he accepted these setbacks with good grace and humour.

Robert Fortune was born on 16 September 1812 at Kelloe in Berwickshire, and after his early education in the local parish church he began an apprenticeship in the nearby gardens of Mr Buchan. Clearly an adept pupil and a quick learner, his first big break came in 1840 when he moved the short distance from the gardens at Moredun near Edinburgh (where he had been working for a few years) to the Botanic Garden in the city. Here he was put to work as a pupil of the capable William McNabb, renowned as a hard taskmaster. In a similar way to David Douglas, Fortune earned the respect and admiration of his superior through his diligence and natural aptitude. Indeed, when in 1842 he applied for the position of Superintendent of the Hothouse Department at the Horticultural Society's garden at Chiswick, London, he got the job primarily on McNabb's recommendation.

Within a few months of taking up his post in London Fortune was selected for the Society's latest plant collecting expedition. The contract that the Horticultural Society required him to sign was, to say the least, demanding. In essence Fortune was to spend a year in China gathering information about Chinese gardening and collecting new plants and seed. He was reminded that hardy plants were of the greatest importance to the Society 'and that the value of plants diminishes as the heat required to cultivate them is increased'. The only exceptions were orchids, aquatics and plants with 'very handsome

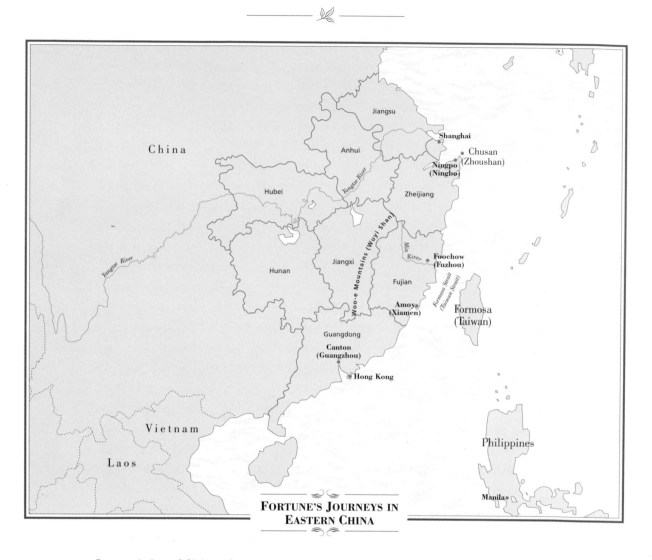

**FORTUNE'S JOURNEYS IN
EASTERN CHINA**

flowers'. In addition, he was to pay particular attention to twenty specific queries listed by the Society. These included keeping a lookout for blue peonies, yellow camellias, double yellow roses, azaleas, lilies, oranges, peaches and various types of tea.

In return the Society was less than generous. Fortune was paid a miserly salary of £100 per annum – the same sum Masson had received some seventy years earlier. Furthermore, although he was expected to run the risk of personal danger, he was offered only a 'life preserver' (a stick weighted with lead) for personal protection. It was only with much persuasion that the Society changed its mind – a decision that was later to save Fortune's life. He was permitted to take a shotgun and a brace of pistols, but somewhat churlishly he was ordered to sell them on his departure from China and return the money to the Society. Perhaps it was no wonder that on his second trip to China Fortune opted to work for the East India Company, particularly given their offer of a 500 per cent pay increase.

With preparations complete, on 26 February 1843 Fortune left Britain aboard the *Emu*, bound for Hong Kong, where he arrived on 6 July after a four-month journey. On the outward journey his Wardian cases (a highly effective portable greenhouse invented by Dr Ward in the late 1830s and used for transporting live plants) were full of plants for the colony, all of which arrived in good health. He found the new British colony to be in a 'deplorable condition' – fever was rife and bands of robbers roamed the streets at night. Not wishing to spend more time than was necessary in such an unhealthy environment, Fortune was soon on the move again. He sailed up the 'barren' coast to the city of Amoy, arriving on 3 September, but if he had hoped to find more favourable conditions he was to be severely disappointed. He wrote disapprovingly in his book *Three Years' Wandering in China* that Amoy:

> is one of the filthiest towns which I have ever seen, either in China or elsewhere; worse even than Shanghae, and that is bad enough. When I was there in the hot autumnal months, the streets, which are only a few feet wide, were thatched over with mats to protect the inhabitants from the sun. At every corner the itinerant cooks and bakers were pursuing their avocations, and disposing of their delicacies; and the odours which met me at every point were of the most disagreeable and suffocating nature.

The general state of dilapidation that Fortune encountered on his travels in China surprised and dismayed him. Having heard so much about the sophistication of the Chinese, it came as a shock to discover a static agricultural society instead. Fortune concluded that:

> There can be no doubt that the Chinese empire arrived at its highest state of perfection many years ago; and since then it has rather been retrograding than advancing. Many of the northern cities, evidently once in the most flourishing condition, are now in a state of decay, or in ruins; the pagodas which crown the distant hills, are crumbling to pieces, and apparently are seldom repaired; the spacious temples are no longer as they used to be in former days.

Fortune made a number of botanical forays inland from Amoy and experienced for the first time the interest aroused by the appearance of a foreigner. Initially the villagers were hostile and shouted threats at him, but Fortune found that by ignoring the commotion and marching resolutely on, the locals would quieten down. Soon he would be surrounded by hundreds of curious people, all eager to discover the purpose of his visit. The only unfriendly creatures left were the local dogs, who had 'a great antipathy to foreigners and will scarcely make friends with them'.

At the end of September 1843 Fortune set sail for the Chusan (now Zhoushan) Archipelago in the north. It was now the monsoon season, and the Formosa (now Taiwan) Strait, which separates China and Taiwan, was being buffeted by northerly gales. Fortune's vessel was caught in a violent storm soon after leaving port and began to plunge alarmingly through the towering waves. At the height of the gale a large fish weighing in excess of 30 lb was thrown out of the sea and crashed through the skylight to land in front of the surprised captain on the poop deck cabin table. Eventually they reached the safety of a local bay, where Fortune transferred to another ship and set out again. They had almost cleared the strait when an even more ferocious gale blew up. The sails were ripped apart by the howling wind, the bulwarks were washed overboard and the terrified crew sought sanctuary beneath the long boat as waves lashed the deck. Soon the boat had been driven back way beyond its starting point. Fortune was down below when suddenly the boat was hit with terrible force. Glass from the skylight rained down around him and seawater began to pour into the cabin. Rushing out into the stormy night, he found that the weather bulwark had been stoved in and the crew and the longboat were hanging precariously on to the other side of the deck. For three days the gale blew the powerless boat about until finally it abated enough for the supply sails to be hoisted and a course set for the nearest land. On inspecting his belongings after the storm, Fortune discovered to his dismay that two Wardian cases containing plants from Amoy had been destroyed.

He was now just 50 miles north of Amoy, but was determined to make the most he could out of adversity and set about exploring the countryside. He was warned that the inhabitants were an unruly lot, but his confidence in his safety was bolstered by his previous encounters outside Amoy. Spying a fine-looking pagoda on top of a nearby hill, Fortune decided to climb up to it in order to survey the lie of the land. As usual he was soon surrounded by several hundred Chinese, who avidly watched him and his servant collect plant specimens as they walked towards the hill. Fortune's silk cravat caused great interest and he was offered various gifts in exchange for it. Not tempted to part with his neckerchief for a handful of chillies, weeds or even a bottle of the local spirit,

Fortune increased his pace in order to leave the crowd behind. On reaching the pagoda he found it to be in a dire state of repair, and having admired the view he decided to make his way back to the ship. Fortune's newly found 'friends' awaited him at the bottom of the hill and began to press ever more closely as he walked along. Suddenly Fortune felt a hand in one of his pockets and turned around to see one of the locals running off with a letter of his. He discovered that he had also been relieved of several more valuable items:

> This incident stopped my progress, and made me look about for my servant, whom I saw at some distance attacked by about eight or ten of the fellows. They had surrounded him, presenting their knives, and threatening to stab him if he offered the least resistance, at the same time endeavouring to rob and strip him of everything of the slightest value, and my poor plants collected with so much care were flying about in all directions.

Fortune rushed towards the crowd, who ran away at his approach, leaving behind his unfortunate servant shaken but unharmed. They quickly picked up the less damaged plant specimens before heading back to the ship in some haste. Among the plants that survived were some fine *Campanula grandiflora* roots and a new species of abelia (*Abelia chinensis*), both of which later arrived safely back in Britain.

Fortune's attitude towards the Chinese varied quite considerably. On the one hand he asserts:

> I am far from having any prejudice against the Chinese people. On the contrary, in many respects they stand high in my estimation. During the last three years I have been continually among them; wandering over and among their hills, dining in their houses, and sleeping in their temples: and from this experience I do not hesitate in pronouncing them a very different race from what they are generally supposed to be.

Indeed he frequently records the acts of kindness shown to him and praises the politeness of many of the locals. However, he describes the southern Chinese as 'being remarkable for their hatred to foreigners and conceited notions of their own importance, besides abounding in characters of the very worst description, who are nothing less than thieves and pirates'. He found the Chinese in the north much more agreeable but, even so, declares that: 'A great proportion of the northern Chinese seem to be in a sleepy or dreaming state, from which it is difficult to awake them.' His experience at the hands of the 'thieves and pirates' undoubtedly coloured his opinion, and his British sense of superiority is just as lamentable as that he found in the Chinese.

Fortune finally reached the Chusan islands after an uneventful ten-day voyage. He was delighted by the rich display of vegetation set among the towering peaks and sloping valleys of the islands (which he claimed reminded him of the Highlands of Scotland). Azaleas of 'dazzling brightness and surpassing beauty' clad the mountains and 'clematises, wild roses, honeysuckles, *Glycine sinensis* and a hundred others, mingle their flowers with them, and make us confess that China is indeed the "central flowery land"'. The tallow tree and camphor tree were abundant, as were forests of bamboo and conifers, including Kerr's Chinese fir (*Cunninghamia lanceolata*). Fortune was enchanted by the beauty of these islands and made frequent visits to them during his sojourn in China.

He now headed west to Ningpo, at the tail end of autumn. It was not an especially profitable visit as far as plant hunting was concerned, but he was able to observe the 'tree dwarfing' practice of bonsai and visited a few of the gardens

A lithograph from *Three Years' Wandering in China,* showing the Treaty Port of Ningpo and junks similar to those that Fortune was travelling on when he was attacked by pirates.

of the Mandarins. The weather grew worse and Fortune began to suffer from the cold. He found the Chinese houses to be extremely inhospitable, as they had large paper windows and were full of crevices through which the biting wind blew remorselessly. Some light relief was found on a journey into Ningpo's hinterland, when Fortune witnessed the unusual sight of men fishing with tame cormorants. The fishermen tied a piece of string around the throats of the birds to prevent them swallowing any of their catch and set them into the water. When a cormorant caught a fish it would docilely swim back to the fisherman and deposit the fish in the boat. Fortune noticed that if a particularly large fish was caught, several other cormorants would help the captor carry its prize back to its master.

By the end of 1843 Fortune had reached Shanghai, a densely populated trading centre on the banks of the mighty Yangtze river, and the most northerly port open to foreigners. Here, as in all other parts of China, strangers were viewed with distrust and dislike. On his walks through the narrow crowded streets Fortune became accustomed to being greeted as 'Kwei-tsz' or Devil's child. This antipathy extended to any business ventures that visiting merchants attempted. Fortune knew that there were a number of nurseries in the local vicinity but had great trouble locating them – the Chinese would either deny they existed or claim they were a great distance away. At last Fortune persuaded some children to direct him to a nursery, but when he approached it the front gate was immediately slammed shut and he had to return the next day with the British Consul to ensure that he was granted access. After a few months' work, the tenacious Scot managed to build up a warm friendship with the Chinese gardeners and obtained a number of fine plant specimens as a reward, including a valuable collection of Moutan or tree paeonies, the elegant Japanese cedar (*Cryptomeria japonica* var. *japonica*) and the delicate Japanese anemone (*Anemone hupehensis* var. *japonica*). These successes went some way to compensate for his miserable living conditions. Here, as at Ningpo, Fortune found the bitter weather almost intolerable and recorded gloomily: 'Our bed-rooms were miserably cold: often, in the morning, we would find ourselves drenched in bed with the rain; and if snow fell, it was blown through the windows and formed "*wreaths*" on the floor.'

Early in 1844 Fortune sailed down to Hong Kong in order to dispatch his plant collection back to England. With time on his hands, he decided to explore the countryside around Canton and nearly lost his life in the process. Walking along a country road between fields and gardens, he was hailed by a soldier on horseback who bade him return the way he had come. Fortune, who did not speak the Cantonese dialect, simply thought the soldier did not want him in the area and marched on regardless. Soon afterwards he was surrounded by 'several groups of ill-looking fellows who seemed to be eyeing me narrowly'. Spotting an enclosed hillside cemetery, he decided to go through its gates in an attempt to lose his companions. The ruse did not work and he began to be jostled by the crowd, who were demanding 'comeshaws' or presents. When he reached the top of the hill, Fortune could see that he was trapped. The Chinese closed in on him, his cap and umbrella were quickly grabbed, and several ruffians began to wrestle him out of his overcoat. Realizing that he was in grave danger, Fortune gathered all his strength and 'threw myself upon those who were below me, and sent several of them rolling down the side of the hill. This, however, was nearly fatal to me, for, owing to the force which I exerted, and the uneven nature of the ground, I stumbled and fell; but fortunately I instantly recovered myself, and renewed the unequal struggle, my object being to reach the door of the cemetery by which I had entered.'

Realizing that the 'Fankwei' or foreign devil was trying to escape, the muggers shouted down to their associates to close the cemetery gates. Fortune fought free of his last assailant and crashed through the gates just as they were closing. The danger, however, had not yet passed, for although he was now on the open road a large crowd had gathered and some began hurling stones. A brick struck him in the middle of the back, forcing him to lean against a wall to recover his breath. He was immediately engulfed by the thieves once again and relieved of more personal possessions. For over a mile Fortune had to alternate between running and fighting with the locals until he finally left their territory. Bruised and suffering from sunstroke as a result of losing his hat and umbrella, he staggered home, grateful to be alive.

During the next year and a half Fortune criss-crossed the Chinese countryside, adding more and more floral trophies to his collection. He became enraptured with the variety and beauty of the scenery and wrote enthusiastically about the hillsides covered with pines, cypresses and junipers, the rich and fertile valleys of tea, tobacco and corn, and the majestic mountain ranges that dominated the landscape. His admiration of the Chinese increased as he became better acquainted with their customs and habits, to the extent that once he had mastered the use of chopsticks he declared: 'In sober truth, they are most useful and sensible things, whatever people may say to the contrary; and I know of no article in use amongst ourselves which could supply their place.'

Although Fortune was a staunch Protestant, he respected the devout and humble Buddhist priests and was always grateful for the friendly welcome he received when he stayed at their temples. On one visit to a temple outside Ningpo, however, he again narrowly escaped an untimely death. In remote areas the priests protected their crops from wild boars by digging deep pits which were half flooded with spring water. They camouflaged the holes with sticks and grass and placed rubbish on top to attract the animals. Although he was warned about the dangers of these pits, Fortune walked blindly on to one during a plant hunting excursion. The ground around the edge of the pit gave way beneath his foot and it was only by grabbing an overhead branch that he managed to stop himself from falling in. On examining the pit he realized that he would have had little chance of being able to climb out and would most probably have died before being discovered. His thoughts turned to David Douglas, who had met his 'melancholy end' in similar circumstances, and he was doubly thankful for his escape. The risk had been worth it, however, for here Fortune found the Japanese snowball (*Viburnum plicatum* 'Sterile') and *Forsythia viridissima*.

In April 1844 Fortune returned to Shanghai, where he collected more tree paeonies. The summer months were spent visiting the Chusan islands, where he found *Weigela florida*: 'certainly one of the most beautiful shrubs of northern China, the *Weigela rosea*, was first discovered in the garden of a Chinese mandarin near the city of Tinghae'. On one occasion he had a further 'narrow escape from a watery grave' while crossing from Ningpo. A strong offshore wind blew up and the crew beseeched Fortune not to put out to sea, but being in a hurry to reach the islands, he ignored their pleas and instructed the captain to set sail. As soon as they reached the open sea it became clear that he had made a dreadful mistake, but the strong tide prevented the little boat from turning back. The captain hoisted the sail, and just as Fortune was questioning the prudence of using so much canvas, a huge wave broke over the side of the boat and filled her from bow to stern with seawater. The situation was now critical, for with the sail still up the boat could roll and capsize at any moment. The crew fortunately got the sail down swiftly and smoothly, and the little craft righted itself. They implored Fortune to turn the boat around, but realizing the danger of such a manoeuvre he ordered the helmsman to steer for the nearest shelter ahead. The sailors, believing the situation hopeless, lost all self-control and began undressing, getting ready for the inevitable swim. Suddenly the wind dropped for a short time and they were able to hoist more sail and reach a small island. As soon as the boat was anchored the crew regained their composure and began bailing out the water: 'We were in a most pitiful condition, all our clothes and beds being completely soaked with sea water; some plants, but luckily only duplicates, which I had with me, were, of course, totally destroyed; but our hearts were light, and we were thankful that our lives had been saved.'

Despite his many close escapes, Fortune's enthusiasm and determination never waned. In June 1844 he decided to visit the forbidden city of Soochow (now Wuhsien), famous for its fine works of art. To do this he simply cut his hair, donned the clothes of a Chinaman and set out. Travelling inland from Shanghai by canal boat, he arrived at the walls of a small town called Cading in the early evening and, after tying the boat up for the night beneath the ramparts, soon fell asleep in his cabin. He was woken up several hours later by a cool breeze blowing through the cabin. On getting up to close the window he discovered that robbers had broken into the boat, stolen all his belongings except his money, which he had prudently hidden under his pillow, and set the boat adrift. Several days later, dressed in a newly acquired Chinese outfit and with a long ponytail attached to his shaven head, Fortune entered the fabled Soochow. Any fears of discovery were soon dispelled when none of the locals paid him the slightest attention as he passed through the city walls.

Although in appearance Soochow was similar to many other towns in northern China, it was clearly far more prosperous than its neighbours. The buildings were in a fine state of repair, the shops were large and doing flourishing trade, and ornamental lakes decorated the landscape. Fortune confirmed that the women lived up to their reputation as the prettiest in the country, although he found their shrunken feet and white-painted faces were not to his taste. He purchased a delightful double yellow rose and a gardenia with large white blossoms from the local nursery, but found little else of interest. On arriving back at Shanghai he was forced to go ashore in his Chinese costume and was highly amused when his British friends failed to recognize him for several minutes.

Fortune now oversaw the loading of his plant collection, divided among four different vessels for safety, for shipment back to the Horticultural Society in London. This consignment included the yellow-flowered winter jasmine (*Jasminum nudiflorum*) and members of the *Rhododendron obtusum* group. In January 1845, wishing for a change of scene, he paid a brief visit to Manila in the Philippines, where he managed to obtain a fine specimen of the orchid *Phalaenopsis amabilis*, along with numerous leech bites for his efforts, before

Fortune's introductions included many shrubs that lighten the winter garden. One of the brightest is the winter-flowering jasmine, *Jasminium nudiflorum.*

heading north again. Much time was spent dashing frantically between collecting sites, gathering duplicate specimens in case anything should happen to the originals on their journey to England, and Fortune often had to take whatever means of transport he could find. At the mouth of the Ningpo river he boarded a small junk heading for Chapu. The other passengers were a motley assortment from all walks of life, although according to Fortune they shared a common dislike for personal hygiene. As they were all huddled together in the central cabin it was virtually impossible for him to avoid catching the fleas that infested his fellow travellers. He built a barricade of his boxes around himself but these were soon knocked over by the rolling of the boat. The Chinese spent much of the night smoking pipes of tobacco and opium:

> When morning dawned, the scene which the cabin presented was a strange one. Nearly all the passengers were sound asleep. They were lying in heaps, here and there, as they had been tossed and wedged by the motion of the vessel during the night. Their features and appearance, as seen in the twilight of a summer morning, were striking to the eye of a foreigner.

From Shanghai, with two companions, Mr Shaw and Captain Freeman, Fortune moved down the coast and sailed up the Min river to the city of Foochow, close to the Woo-e (now Wuyi Shan) Mountains. The journey upriver provided some of the most spectacular scenery encountered so far. Low flat plains gave way to jagged mountains, which fell into the river itself. The fertile hillsides were terraced with crops of sweet potatoes and peanuts, and the thickly wooded river banks were studded with numerous temples. Springs forced their way through the patches of barren granite and fell in long cascades to the water below. Fortune concluded that 'viewing the scenery as a whole – the beautiful river, winding its way between mountains, its islands, its temples, its villages and fortresses – I think, although not the richest, it is the most romantic and beautiful part of the country which has come under my observation'.

After such a glorious approach, Foochow could only be a disappointment. To make matters worse, heavy rain had caused the streets to flood (up to chest-height in places), and the residents took exception to the presence of foreigners. They harried their porters and threw buckets of water over the sedan-chairs in which Fortune and his associates were travelling. Staying only long enough to collect a few new plants from the local nursery and to examine the countryside, Fortune was soon ready to depart again. He obtained a berth on a junk carrying a cargo of timber to Ningpo, and sailed down to the mouth of the Min river. Here the junk joined a convoy of 170 other boats all heading up the coast. As a precaution against being caught in a pirate attack, the unarmed junks always sailed together whenever possible. Before they set out, Fortune was struck down by a severe fever and lay sweating and shivering uncontrollably in his cot for several days, his moments of consciousness spent considering the prospects of being buried in a lonely spot on the banks of the Min river.

The convoy set off together but within hours had separated into little groups of three and four. Late that afternoon the captain of the junk appeared agitatedly beside Fortune's bed to inform him that five 'Jan-dous' or pirate boats had been spotted lying in wait for them. Although he did not take the threat seriously, Fortune crawled out of his cot, loaded his twelve-bore shotgun and pistols and went up on deck. One look through a small telescope confirmed the captain's worst fears – the junks bearing down on them were filled with armed men. The sailors on Fortune's boat ran around in a state of panic. Some tried to hide their money under the cabin floorboards and in the ballast, while others changed into rags in order to fool the pirates into thinking they were worthless serfs. Fortune held out little hope that he could fight off all five vessels and did not expect to get much assistance from the terrified crew, whose only means of defence was to throw small stones from the ballast. However, he was determined to make a spirited stand, for he knew that any Westerners caught by pirates were invariably killed. At a distance of 200 or 300 yards the nearest junk fired a broadside.

> All was now dismay and consternation on board our junk, and every man ran below except two who were at the helm. I expected every moment that these also would leave their post; and then we should have been an easy prey to the pirates. 'My gun is nearer you than those of the Jan-dous,' said I to the two men; 'and if you move from the helm, depend upon it I will shoot you.' The poor fellows looked very uncomfortable, but I suppose thought they had better stand the fire of the pirates than mine, and kept at their post. Large boards, heaps of old clothes, mats, and things of that sort which were at hand, were thrown up to protect us from the shot; and as we had every stitch of sail set, and a fair wind, we were going through the water at the rate of seven or eight miles an hour.

Two more shots were fired, the second flying harmlessly above the heads of the helmsmen and past the sails, as the pirate ship sailed closer and closer. The pirates screamed triumphantly as they re-loaded their weapons for the final attack. Fortune later wrote: 'Their fearful yells seem to be ringing in my ears even now, after this lapse of time, and when I am on the other side of the globe.' He had one desperate plan to save the junk. He knew that in order to deliver a broadside the helm had to be drawn down, and by watching the helmsman's actions he could predict when the pirates were about to fire. Instructing the two terrified sailors next to him to copy his movements, Fortune dived to the floor as the pirate helmsman swung the boat around. A volley of shots exploded above his head but missed everyone on deck. With the pirates now no more than 20 yards away, Fortune jumped to his feet and fired both barrels of his shotgun into the jeering hordes.

Had a thunder-bolt fallen amongst them, they could not have been more surprised. Doubtless, many were wounded, and probably some killed. At all events, the whole of the crew, not fewer than forty or fifty men, who, a moment before, crowded the deck, disappeared in a marvellous manner; sheltering themselves behind the bulwarks, or lying flat on their faces. They were so completely taken by surprise, that their junk was left without a helmsman; her sails flapped in the wind; and, as we were still carrying all sail and keeping on our right course, they were soon left a considerable way astern.

The danger was far from over, however, as the second pirate boat now descended on the junk, firing broadsides. Fortune's two companions begged and pleaded with him to return fire, but again the sturdy Scot waited until the last moment before blasting the pirates with his shot. This time Fortune intentionally aimed at the enemy helmsman, and with the man cut down the pirate junk lost its momentum and fell back. When those on the remaining junks saw what had happened to their partners in crime, they lost their appetite for a fight and gave up the chase. Fortune's 'heroical companions' now left their hiding places and rushed up on deck to taunt the fast-disappearing pirates. All on board began singing the praises of their foreign saviour and some of the crew actually knelt down at his feet in an act of reverence. Feeling exhausted and still sick with fever, Fortune retired to the comfort of his cot.

Two days later the captain again rushed down to Fortune's bedside, this time to inform him that six pirate vessels had been sighted ahead. With more time to prepare for this attack, Fortune decided to try to fool the enemy into thinking they were more heavily armed than they really were. Knowing that foreigners and their weapons were greatly feared, he dressed the least Chinese-looking members of the crew in all his spare clothes and told them to brandish the short levers used for hoisting the sails, which, from a distance, looked like firearms. As soon as the nearest junk fired a broadside, however, all the crew on deck immediately ran for cover and once again Fortune had to threaten the helmsman to keep him at his post. Fortune carried out his tactic of returning fire at the last minute with the same success as before. Two more pirate boats came up and fired some shots at the junk but they did not attempt a full assault. As darkness fell and the pirate boats passed out of sight, Fortune began to feel unwell. 'The fever, which I had scarcely felt during all this excitement, now returned with greater violence, and I was heartily glad to go below and turn into my bed.'

Now that the danger had passed and they were in safe waters, the captain and his crew showed a complete lack of gratitude towards Fortune. In return for defending the junk he had been promised that he would be dropped at Chusan harbour. With only a few miles to go, the captain nonchalantly informed him that they were going straight on to Ningpo. Incensed by the unthankful captain's behaviour, Fortune informed him that Englishmen never allowed promises to be broken and that he would stand by the helmsman with his gun and would not be responsible for his actions if they steered off the course to Chusan. They duly arrived at Chusan, and Fortune, eager to be reunited with his plant collection at Shanghai, crawled on to an English vessel bound for the port. After recovering from his fever he sailed down to Hong Kong, where he dispatched eight Wardian cases full of plants back to England. These included

---

The sweet scent of the winter-flowering honeysuckle, *Lonicera fragrantissima* fills the air throughout the coldest months.

three winter-flowering shrubs, *Mahonia japonica*, *Lonicera fragrantissima* and *L. standishii*, the lace-bark pine (*Pinus bungeana*) and the graceful bleeding heart (*Dicentra spectabilis*). Fortune himself sailed from Canton on 22 December with a further eighteen cases of plants and arrived back at London on 6 May 1846.

In August 1848 Fortune returned to China, this time sent by the East India Company 'for the purpose of obtaining the finest varieties of the Tea-plant, as well as native manufacturers and implements, for the Government plantations in the Himalayas'. Donning his disguise as a Chinese man, he hired an interpreter and set off for the Hwuy chow district, 200 miles inland from Ningpo. He travelled by boat or sedan-chair and stayed at rudimentary Chinese inns or in Buddhist temples. Although there were none of the hair-raising adventures of the first trip, there was plenty of opportunity for some profitable plant collecting. One day when Fortune and his assistant were out gathering plants he happened to look up and see a new type of cypress (now known as the mourning cypress, *Cupressus funebris*) nearby. In a great state of excitement he hurried over to gather a seed sample, only to discover that the tree was growing in the walled garden of an inn. He briefly considered climbing over the wall but remembered 'that I was acting Chinaman, and that such a proceeding would have been very indecorous, to say the least of it'. Not to be denied his new-found treasure, Fortune decided to try a more subtle approach:

> We now walked into the inn, and, seating ourselves quietly down at one of the tables, ordered some dinner to be brought to us. When we had taken our meal we lighted our Chinese pipes, and sauntered out, accompanied by our polite host, into the garden where the real attraction lay. 'What a fine tree this is of yours! we have never seen it in the countries near the sea where we come from; pray give us some of its seeds.' 'It is a fine tree,' said the man, who was evidently much pleased with our admiration of it, and readily complied with our request. These seeds were carefully treasured; and as they got home safely, and are now growing in England, we may expect in a few years to see a new and striking feature produced upon our landscape by this lovely tree.

Fortune could have saved himself the cost of a meal, for shortly afterwards he spotted another mourning cypress growing in a dilapidated old garden.

Left: The lace-bark pine (*Pinus bungeana*), introduced by Fortune in 1846, is one of the most elegant of the Oriental pines. It has peculiar bark, rather like that of a eucalyptus.

Deciding that boldness was the best policy, he and his servant marched through the gates 'with the coolness of Chinamen' and picked up some ripe seeds:

> Having taken a survey of the place, we were making our way out, when an extraordinary plant, growing in a secluded part of the garden, met my eye. When I got near it I found that it was a very fine evergreen Berberis, belonging to the section of Mahonias ... It had but one fault, and that was, that it was too large to move and bring away. I secured a leaf, however, and marked the spot where it grew, in order to secure some cuttings of it on my return from the interior.

Fortune successfully collected tea plants from the Hwuy-chow district and the Chekiang province. He also gathered specimens from the Ningpo district, Chusan and the Woo-e mountains, and supervised the transfer of 23,892 young plants and approximately 17,000 seedlings, along with eight Chinese tea growers and their equipment, to the foothills of the Himalayas. Tea plantations were established in Assam and Sikkim and it became one of northern India's principal exports during the second half of the nineteenth century. Tea's importance is shown by the value of imports into Britain, which rose a staggering 837 per cent in the seventy-five years between 1854 and 1929 (£24,000 to £200,880).

Fortune made two further trips to China. The first (1853–6) was disrupted by the Tai-ping Revolution, but despite being robbed and prevented from exploring, he found the lovely *Rhododendron fortunei*. His often forgotten fourth trip (1858–9) was on behalf of the Government of the United States of America, which wished to establish its own tea industry. As a result of his explorations 32,000 plants were grown, but the American Civil War effectively put paid to the plan.

For his fifth trip (1860–62) Fortune once more took advantage of political changes and switched his destination to Japan. This trip was a private, speculative expedition. The story of Japan's closure to the West is similar to that of China. In 1639 the Japanese authorities banished all missionaries, and their attitude towards foreigners became even less accommodating than the Chinese in the years before the Opium Wars. In 1853 the United States Expedition to the China Seas and Japan set out with the aim of establishing diplomatic and trade links. The first plant hunters to begin the extensive exploration were the Americans S. Wells Williams (1812–84) and Dr James Morrow (1820–65). When the Treaty of Kamagawa was signed on 31 March 1854, the two men

'View in the Green Tea District' from Fortune's book *A Journey to the Tea Countries of China*. Fortune was responsible for the establishment of the East India Company's tea industry in India.

wasted no time in 'going bush'. Their collections were herbarium specimens, however, and it was left to a keen amateur to send back the first significant collection of live Japanese plants to America. In 1861 Dr George Rogers Hall sent a unique collection of plants in Wardian cases to a friend, Francis L. Lee of Boston. Lee, who was about to enlist in the Union Army in the Civil War, entrusted the plants to his friend, the eminent horticulturist Francis Parkman. The cases of treasure that inadvertently came into Parkman's possession contained what have become some of the most popular garden plants on both sides of the Atlantic. The conifers included the Japanese yew (*Taxus cuspidata*), *Chamaecyparis pisifera* and ten garden forms of *C. obtusa*. There were also three lovely magnolias, *M. stellata*, *M. kobus* and *M. halleana*, the dogwood *Cornus kousa*, *Wisteria floribunda* and various forms of *Acer palmatum*.

Fortune had a much easier time travelling in Japan than in China. Despite the threat from the feudal princes or *daimyos*, who saw it as their sworn duty to massacre foreigners at every opportunity, he avoided any unpleasant incidents. His collections included a range of vibrant chrysanthemums and a lovely variegated bamboo (*Pleioblastus variegatus*). With its usual entrepreneurial flair, Veitch & Sons, the most influential of all the nineteenth-century nursery companies, was quick to react to the new situation in Japan, and in 1860 John Gould Veitch arrived at Nagasaki. He was just in time to join a party of Englishmen led by the British Envoy, Sir Rutherford Alcock, who became the first Europeans to climb the sacred Mount Fujiyama. Reaching the summit on 11 September 1860, the group gave a stereotypical display of 'Victorian Englishmen abroad'. They drank the health of Her Most Gracious Majesty the Queen, hoisted the Union Jack on a makeshift flagpole, fired a twenty-one-gun salute with their revolvers and shouted 'God Save the Queen!'

In terms of Japanese plant introductions, John Veitch was more successful than Fortune. In only four months he collected seventeen new conifers, including *Larix kaempferi*, *Picea bicolor*, *Pinus thunbergii*, *P. parviflora*, *Juniperus chinensis* 'Aurea', *Chamaecyparis obtusa* and *C. pisifera* 'Squarrosa', as well as *Magnolia liliiflora* 'Nigra' and *Lilium auratum*. However, since the collections of Fortune and Veitch were sent from Japan aboard the same ship, it became a contentious issue as to who got the glory for introducing certain plants. George Hall did not have this problem, and his 1862 collection, which he brought back with him to Rhode Island, further established him as the father of Japanese plants in America. Whether Hall's first consignment got back before that of Fortune and Veitch is another question!

Fortune arrived back in Britain in January 1862 and settled in Kensington, London. John Veitch, meanwhile, travelled from Japan to the Philippines in search of *Phalaenopsis* orchids before returning home in 1863. While Fortune enjoyed eighteen years of comfortable retirement, paid for by the success of his books and the sale of oriental antiques, which he had collected on his travels, Veitch fell victim to tuberculosis, dying prematurely at the age of thirty-one.

Robert Fortune's most significant legacy was the transfer of tea plants from China to India. From a gardener's perspective he is credited with discovering over 120 new species, and as one French admirer put it, 'all Europe is obliged to him'. He was the first plant hunter to bring back a wide selection of Chinese and Japanese plants, thus awakening the public to the diversity and beauty of the Oriental flora. Every part of the garden was graced by Fortune's work, since he brought back shrubs, trees, herbaceous perennials and bulbs. However, an analysis of his plant introduction list indicates that the winter display benefited the most, with the addition of shrubs such as *Jasminum nudiflorum*, *Lonicera fragrantissima*, two forsythias (*F. viridissima* and *F. suspensa* var. *fortunei*) and three mahonias (*M. japonica*, *M. japonica bealei* and *M. fortuneii*). These were augmented by his conifers, for example *Cryptomeria japonica*, *Cupressus funebris* and *Chamaecyparis pisifera*, which were also added to the pinetum and arboretum. The shrubbery, which enhanced the spring and summer display, benefited from finds such as *Weigela florida*, *Abelia chinensis*, *Viburnum plicatum* 'Sterile' and the herbaceous *Anemone hupehensis* var. *japonica*. In 1878 another way to display plants from the Orient was recommended – the 'Japanese garden', a theme garden containing a geographical collection of plants. Although the idea cannot be said to have directly changed garden fashions, this was another development in the Victorian landscape.

Robert Fortune's Plant Introductions

The date beside each plant name is the date of its introduction into Britain.

Cryptomeria japonica (1842)

Cryptomeria = (Gk) *krypto*, to hide; *meris*, with flower parts hidden

japonica = (Lat.) Japanese

Dense spirals of slightly bent, small, bright green leaves crowd along the slender branchlets, contrasting with the lovely reddish, peeling bark. The Japanese cedar is a large, elegant, fast-growing columnar tree, somewhat similar to the North American redwoods. Thomas Lobb introduced two Japanese forms from the Buitenzorg Botanical Gardens, Java: the lovely 'Elegans' (1854), with billowing masses of feathery, bluish-green juvenile foliage, which colours to red-bronze in autumn, and 'Lobbii' (1853), an attractive conical tree with open tufted growth.

Native to Japan and China, but its exact natural distribution is obscure. It is an important timber tree in Japan, reaching 180ft (55m).

Anemone hupehensis var. *japonica* (1844)

Anemone = from (Gk) *anemos*, wind; *mone*, habitation

hupehensis = (Lat.) of Hupeh, China

japonica = (Lat.) Japanese

Charming, clear pink, five-petalled flowers, each with a prominent centre of bright yellow stamens, hover airily on narrow 24–50in (60–130cm) branched stems from August to October. A lovely spreading perennial with handsome clumps of vine-like dark green foliage. Good selections include the very free-flowering 'September Charm' (1932).

Occurs in eastern China and has been long cultivated there and in Japan. The species, *A. hupehensis*, ranges from Hubei to Sichuan and Yunnan in China, growing in open, rocky sites, to shady scrub and cliffs at 1,950–8,200ft (600–2,500m).

Fortune often found *Anemone hupehensis* var. *japonica* growing on graves, 'a most appropriate ornament to the last resting-place of the dead'. It is a wonderful sight planted *en masse*, especially when the flowers are nodding in a gentle breeze.

Forsythia viridissima (1844) and
Forsythia suspensa var. *fortunei* (1860)
Forsythia = after William Forsyth, Superintendent
   of Royal Gardens at Kensington Palace
viridissima = (Lat.) very green
suspensa = (Lat.) hanging
fortunei = after Robert Fortune

Wreathed with bright golden-yellow flowers in March to April, the bare branches of these familiar harbingers of spring are followed by deep green leaves, which colour yellow in autumn. *F. viridissima* is an erect shrub and the first to arrive in Britain. *F. suspensa* var. *fortunei* is a superior, vigorous form of the lax, graceful species, introduced in 1850 via Holland (1833). The vigorous *F* × *intermedia* (*F. suspensa* × *F. viridissima*, *c.* 1880, Germany) has produced some fine selections, e.g. 'Spectabilis', 'Lynwood' and 'Minigold'.

Both occur in China, with *F. suspensa* common in west Hubei, growing in scrub and on cliffs at 1,000– 3,900ft (300–1,200m) and reaching 8–10ft (2.5–3m) tall.

Jasminum nudiflorum (1844)
Jasminum = Latin version of the Persian *yasmin*
nudiflorum = (Lat.) flowers before leaves

Sulphur-yellow, delicate flowers decorate bare green stems from November to February. The winter jasmine is a beautiful deciduous shrub of spreading, sprawling habit, with whip-like angular green stems and glossy green leaves. *J. n.* 'Aureum' is a yellow-variegated form (1889).

Ranges throughout western China and has been cultivated there for centuries, reaching 10ft (3m).

Weigela florida (1845)
Weigela = after C. E. Weigel, German botanist
florida = free-flowering

With masses of funnel-shaped rose-pink to reddish flowers in May and June, this medium-sized, spreading deciduous shrub with bright green leaves is the parent of many hybrids. Cultivars include *W. f.* 'Foliis Purpurea' (a dwarf purple-leafed form),

'Gustav Mallet' (free-flowering, deep red), 'Variegata' (cream-edged leaves, pink), and the hybrid 'Bristol Ruby' (vigorous, ruby-red).

Occurs in Kyushu in Japan, Korea and north-east China, growing in scrub and reaching 10ft (3m) tall.

Mahonia japonica (1846) and
Mahonia bealei (1849)
Mahonia = after Bernard M'Mahon
japonica = (Lat.) Japanese
bealei = after Mr Beale of Shanghai

Magnificent, architectural pinnate leaves divided into spine-tipped blue-green leaflets on rounded evergreen shrubs. *M. japonica* bears lax, pendulous racemes of sweetly scented lemon-yellow flowers in November to March, while *M. bealei* produces shorter, almost erect flower spikes. Their nomenclature has been much confused in the past. Fortune also introduced the autumn-flowering *M. fortunei* (1846).

M. bealei is native to woods in Hupeh, Hubei and Sichuan in China and in Taiwan, growing at around 6,500ft (2,000m). *M. japonica* has an uncertain distribution in China but has been long cultivated there and in Japan. It is parent to the excellent hybrid *M.* × *media* (× *M. lomariifolia*) and its cultivars, e.g. 'Charity' and 'Winter Sun'.

Rhododendron fortunei (1855)
Rhododendron = (Gk) *rhodo*, rose; *dendron*, tree
fortunei = after Robert Fortune

Lovely, fragrant funnel-shaped pale lilac-pink to pink flowers in loose trusses of 6–12 appear in May above dark green foliage, glaucous green beneath. New foliage appears with the flowers. The first hardy Chinese species to arrive, and parent of many hybrids. Hybrids include the superb Loderi Group (1901).

Occurs in mountains in south-east China from Kiangsi to Hupeh, Kwangtung and Kiukang at 1,950–2,950ft (600–900m), frequently in deforested areas and by rivers, growing to 28ft (9m).

6

BROTHERS IN THE NURSERY

The Lobbs and the Veitch Dynasty

William Lobb (1809–64); Thomas Lobb (1811–94)

To SET THE SCENE FOR THE NEXT PLANT HUNTERS WE MUST BRIEFLY review the socio-economic changes that had taken place in Britain and discuss how these engendered a new phenomenon – suburbia, with its 'middle-class *nouveaux riches*'. In 1851, the same year that Sir Joseph Hooker returned from Sikkim, Britain, brimming with national pride, decided to show herself off to the world. The form that this pageant took was the Great Exhibition. Held in Hyde Park, it had as its centrepiece the extravagant Crystal Palace, an enormous greenhouse four times the size of St Peter's in Rome. The reasons behind the exhibition were threefold: Britain wished, first, to demonstrate her stability, to show that, unlike much of the rest of Europe, she was at peace; second, with her Empire expanding and relatively secure, to show that she was now consolidating her position as the world's most powerful nation; and third, having recovered from an economic depression that dominated the 1840s, to confirm her refound economic prowess.

Left: From the humid jungles of south-east Asia Thomas Lobb collected the bizarre, carnivorous *Nepenthes* spp., which became popular in Victorian stovehouses. The elegant, fluid-filled pitchers capture and digest anything small enough to fall in.

Above: A detail from the 1886 seed catalogue published by James Veitch's Royal Exotic Nursery, London. The nursery was well supplied by Veitch's nurseries outside the capital.

It was a boom in the key industries of coal, iron and cotton that earned Britain the title 'workshop of the world', and this industrial expansion was inescapably linked with technological advances. Kept a closely guarded secret from other developed countries, these improved techniques meant Britain almost had a monopoly on mass production. Supply was met by demand, both at home and by the almost endless export opportunities offered by the Empire. The Empire was also equally important for its substantial supplies of both cheap raw materials and labour.

It was a prosperous and optimistic Britain that entered the second half of the nineteenth century, and just as the economic scene experienced great change, so did the physical landscape. Industrial growth had already meant the start of mass urbanization; the 1851 census revealed that over 50 per cent of the population of 22.3 million lived in cities with over 50,000 inhabitants. Britain was now indisputably 'the world's first urban society'. As well as a buoyant market for speculative housing developers, the cities created a demand for agricultural produce and helped to create the conditions for the 'Golden Age of Agriculture'. Other factors that stimulated this particular boom were new and improved farming practices and techniques, such as field drainage, the use of fertilizers, the production of high quality seed and mechanization.

Horticulture also benefited from technological advances and labour-saving devices, which made gardening easier. In 1832 the first lawn mower was patented by Mr Budding, and it quickly replaced the scythe. This manoeuvrable machine made it easy to cut the grass between the complex flowerbeds that were beginning to stud the lawn. The quick and neat cut produced by Budding's cylinder mower meant that the lawn became an increasingly popular feature, particularly in suburban gardens. Fertilizers and manures were another area of great debate, and an enormous range of materials were 'guaranteed' as certain ways to improve plant growth. Examples included 'Poudrette' (nightsoil mixed with coal ash), chopped animal entrails and even huge quantities of sticklebacks scattered over the soil. Perhaps the most commonly advertised was guano, which after its discovery in Peru in 1842, went on to become an enormous industry. Pesticides were another field for experimentation, and even in today's world of highly toxic organophosphates, to look back at some of the chemicals used sends a shiver down the spine. Pity the poor gardeners who had to apply arsenic, nicotine or soft soap pounded with mercury. There seemed little concern that the chemicals that destroyed pests might also damage human health.

In addition to horticultural improvements, there was a shift in the supply of exotic plants. Up until now, the new species brought back by the plant hunters had gone first either to scientific organizations such as Kew or to a select few who had the wealth and enthusiasm to co-sponsor private expeditions. But, just as J.C. Loudon (see pages 73–4) had identified a commercial opportunity for new publications telling these gardening initiates how-to-grow, many nurserymen now saw a potential market supplying what-to-grow to the ever-increasing number of gardeners. For the first time, many exotics became available to the public at a reasonable cost. This was possible for two reasons. First, plant propagation techniques improved, which meant new exotics could be produced in large numbers; second, the larger nurseries sent out their own plant hunters to gather commercially viable quantities of seed from the wild.

The Veitch dynasty had been founded in 1808, when John Veitch (1752–1839), a Scot from Jedburgh who had moved south to work as a land steward for Sir Thomas Ackland at Killerton House, Devon, rented land at nearby Lower Budlake and established a nursery selling mainly trees and shrubs. The business flourished, and he rented additional land in 1810 before leaving the site in 1832 for larger premises at Mount Radford, near Exeter. This became the famous 'Exeter Nursery'. John went into partnership with his son James (1792–1863) and later with his grandsons, James junior (1815–69) and Robert (1823–85). When he was eighteen years old James junior was sent to train in London, where he spent a year at Mr Chandler's nursery at Vauxhall and a second at Messrs Rollisson of Tooting. On his return to Devon he set about improving and expanding the Exeter Nursery, becoming a partner in 1838. He soon realized that Veitch & Sons could not compete effectively with the large London nurseries from such a distance, and in 1853 he acquired the Royal Exotic Nursery business of Messrs Knight and Perry, renting and later buying their land on the Kings Road, Chelsea. The Kings Road establishment expanded rapidly as James developed an exceptional collection of exotic green-house plants.

Eventually it became unfeasible to run both businesses side by side, and in 1863 they became independent. In Exeter James senior was succeeded by Robert, and this branch of the family business became Robert Veitch & Sons, with Robert succeeded by Peter (1850–1929). The London branch took the name James Veitch & Sons, and here James junior was followed by John Gould (1839–70), Harry James (1840–1924) and Arthur (1844–80). At its zenith the Chelsea nursery was perhaps the largest of its kind in Europe. It was subdivided into eleven departments – Orchid, Fern, New Plant, Decorative, Tropical, Softwood, Hardwood, Vine, Propagating, Seed and Glass – each with its own foreman. Additional nursery sites were added at Feltham (garden plants, florists' flowers and seed production), Langley (tree and bush fruits and later on orchids) and Coombe Wood (hardy plants, rhododendrons and azaleas) to supply the main site.

The impact that the Veitch dynasty had on horticulture is worthy of a book in its own right, but for the purposes of this story their key role was the dispatch of twenty-two plant hunters to collect exclusively for the Veitch nurseries. This far-sightedness proved to be very profitable. James H. Veitch wrote in *Hortus Veitchii* (1907): 'For nearly half a century, however, that spirit of private enterprise had, except in a few instances, given way to the united efforts of corporate bodies and government officials; and it was not till the bold and energetic course which has been pursued by a provincial nurseryman of England was adopted, that a new era in botanical discovery was begun which has placed the name of "Veitch of Exeter" among the worthies of science in our times.'

The most influential collectors were the penultimate, Ernest Wilson, and the first, the brothers William and Thomas Lobb. Few records of the lives and work of these two Cornishmen exist, but it is known that they spent their early years in Egloshayle, where their father, John, was an estate carpenter at nearby Pencarrow, a notable garden developed by Sir William Molesworth. John had a love of natural history (something that both boys clearly inherited), and when a financial crisis hit the family he asked Mr Carlyon, the vicar of Egloshayle, for his advice. Carlyon used John's knowledge of the outdoors to help secure him a position as gamekeeper at Carclew, the home of Sir Charles Lemon. Sir Charles, a relation of the Tremayne family of Heligan, was one of the very first people to receive and grow rhododendron seed from Hooker's Himalayan expedition, having had the seed sent directly to him from India by Sir Joseph Hooker himself.

The early employment of both brothers is somewhat confused, with various accounts offering differing facts. However, it seems that following a good basic education at Wadebridge both brothers worked in the stove houses at Carclew, where Sir Charles, a kindly employer, encouraged the boys in their study of horticulture and botany. Thomas did not remain there long, joining Veitch at Killerton in about 1830, when he was only thirteen. It is implied in *Hortus Veitchii* that William was also working for the Veitch family at this time, but this

The Veitch dynasty was the most dominant force in the British nursery trade for over half a century. Part of the success was due to their far-sighted policy of employing its own plant hunters.

remains unsubstantiated. Whatever the truth, it is a fact that once the nursery moved to Exeter in 1832 John Veitch's thoughts turned to employing his own plant hunters to gather exotics exclusively for the nursery. This was sound business acumen, for if a plant hunter sent back bulk quantities of seed of species already introduced but still rare, this would enable the nursery to sell exotics at an affordable price. Also, if the expedition were carefully targeted, there was the strong possibility that the plant hunter would send home valuable new species that would both command a premium price from the gardening élite and offer scope for plant breeding, with any successful hybrids also commanding a high price.

By the late 1830s William Lobb wished to travel abroad, and Thomas suggested his name to John Veitch, who, impressed with William's botanical knowledge and eye for a good plant, employed him as the nursery's first plant hunter. Veitch had been in contact with Sir William Hooker about a suitable location for an expedition, and South America emerged as the preferred destination. On 7 November 1840, therefore, William Lobb, aged thirty-one, set sail from Falmouth aboard HM Packet *Seagull*, bound for Rio de Janeiro.

Veitch, in contrast to other employers, always ensured that its plant hunters travelled in comfort and were never 'cramped for funds', and William set out with between £300 and £400 per annum available to draw on in various large cities on his planned itinerary. Unfortunately, Veitch did not specify that their collectors keep a journal, so the details about the adventures encountered by the Lobbs and later plant hunters are somewhat sketchy. More unfortunate still is that although the Veitch plant hunters were all required to write back frequently, the Veitch archives have since disappeared. From John's correspondence with Sir William Hooker at Glasgow, and later at Kew, it is known that William Lobb spent the southern winter of 1841 in Brazil and around Buenos Aires in Argentina. He explored the Organ Mountains and discovered *Begonia coccinea*, *Passiflora actinea* and a swan orchid (*Cynoches pentadactylon*) before journeying to Chile via Mendoza and the Upsallata Pass over the Andes. This gruelling trek was well worth the hardship to William – not only did he see the outstanding Andean landscape but he was also saved the perilous sea journey around Cape Horn.

Several recent commentaries have William rediscovering the monkey puzzle tree during his crossing of the Upsallata Pass, but this is contradicted by *Hortus Veitchii*, which states that from Concepción in Chile, 'Continuing his journey southwards, Lobb penetrated the great Araucaria forests, where he

Forests of monkey puzzle (*Araucaria araucana*) stretch out below the brooding, snowcapped Llaima volcano in Chile. *Araucaria* nuts were once the staple food of the indigenous Araucana people.

collected a large quantity of seeds of *Araucaria imbricata* [now *araucana*] and was thus instrumental in bringing this remarkable conifer into general use for ornamental planting.' Furthermore, *Araucaria araucana* occurs naturally only below the latitude 37° south, and the Upsallata Pass is some 300 miles north of this. Therefore it is safe to assume that Veitch was right and that it was on a trip to the south of Chile that William visited the monkey puzzle forests to collect seed.

It is an easy scene to imagine. William, recovered from the ordeal of his Andean crossing, leaves the mild, Mediterranean climate of Valparaiso and travels south by steamer to Concepción, his destination the great forests of the

La Araucana region. It is late summer when he crosses the fertile farmland with fields of ripening crops and vineyards irrigated by meltwater from the towering Andes. As he journeys on into the forests that cloaked the foothills he is accompanied by the ceaseless drone of the Patagonian summer winds, which rustle the leaves of the emerald-green southern beech (*Nothofagus* species).

It is high on the distant mountain slopes that William hopes to reap his reward, leaving the small hillside village at first light with his guide, porters and mules. The path is dry and well-trodden and after a few hours walking they move out of the forest and on to an exposed ridge at 5,250ft. From this vantage he can admire the endless succession of mighty snow-capped mountains and volcanic cones stretching north and south as far as the eye could see. Turning his glance upwards across the chasm, high above on the flanks of the volcano he sees his goal – the impressive stands of monkey puzzle.

Forgetting his weariness, William marches onwards and upwards. Eventually he crosses an open meadow and enters a surreal forest. Surrounding him are tall, perfectly straight cylindrical trunks covered in coarse bark like an elephant's hide, crowned with spidery branches of stiff, pointed leaves. The trees create a silhouette like a giant umbrella, hence the local Spanish name of *paragua*, and to his great relief the huge cones are still heavy with seed. It is now late in the afternoon and the party sets up camp near the trees. William sits and watches the sun set through the monkey puzzle forest, glistening on the lakes below and turning the snow-capped peaks first orange and then pink before they become ghostly white shadows in the cold, clear night air. The following morning everyone busily gathers fallen nuts and William shoots prize cones down from the tree. By midday hundreds of healthy seeds have been gathered and packed, and a very satisfied William returns to the village.

Back in Valparaiso, William dispatched the seed to Veitch & Sons. This first package contained '3,000 seeds', and by 1843 Veitch was offering seedlings at £10 per 100. This new tree, with its extremely graceful juvenile form and highly unusual foliage, was an instant hit, equally popular as an addition to the landscape or as the central focus to a small suburban bedding display. As far as Veitch & Sons was concerned, this single, early reintroduction confirmed that the policy of sending out plant hunters was a commercial success.

Right: The monkey puzzle tree (*Araucaria araucana*) was named by a visitor to the gardens at Pencarrow, Cornwall who, observing a young tree for the first time, commented that the sharp leaves would puzzle any monkey trying to climb it.

William now headed north, and among his Chilean discoveries were *Desfontainea spinosa*, *Mandevilla splendens*, *Hindsea violaceae* and *Tropaeolum azureum*. From Chile he travelled to Peru via Ecuador and southern Colombia. *En route* he collected a number of tender species including *Calceolaria amplexicaulis*, whose pale yellow flowers made it an early favourite for bedding schemes, and the passion flower, *Passiflora mollissima*, which thrives in a cool greenhouse. He set sail from Panama and arrived back in Cornwall in 1844. The trip was judged to be so successful that John Veitch insisted he return to Chile, so in 1845 he set sail for another three-year trip. Once again he broke his journey in Brazil and visited the Organ Mountains. With memories of his overland crossing still fresh in his mind, this time William opted for the sea passage around the Cape to Valparaiso. Instructed to focus on hardy and half-hardy shrubs, he ventured as far south as Valdivia in northern Patagonia. Although we do not have William's own words, a later collector for Veitch & Sons, Richard Pearce, paints a vivid portrait of this area's dramatic scenery:

> It is of the most charming description – gently undulating meadows covered with a carpet of short grass, placid lakes reflecting from smooth surface the mountains around, foaming cataracts and gentle rivulets, deep gorges and frightful precipices, over which tumble numerous dark, picturesque waterfalls reaching the bottom in a cloud of spray, high rocky pinnacles and lofty peaks, surround one on every side.
>
> Nor is the vegetation less beautiful and interesting. At an elevation of 4,000ft the vegetation exhibits a totally different character from that of the coast. Here one finds Antarctic Beeches (*Fagus antarctica* and *F. betuloides*) which constitute with *Fitzroya patagonica* the large forest trees. The *Embothrium coccineum*, *Desfontainea spinosa*, *Philesia buxifolia*, three species of *Berberis*, *Pernettya* and *Gaultheria* are the most abundant of the flowering shrubs whilst the numerous pretty little rock-plants meet one at every step with their various forms and colours.

William also spent time on Chiloe Island, where among his finds were *Berberis darwinii* and *Escallonia macrantha* var. *rubra*. From the mainland came the Chilean fire bush (*Embothrium coccineum*), *Crinodendron hookerianum*, the Chilean bellflower (*Lapageria rosea*), the flame nasturtium (*Tropaeolum speciosum*), three myrtles (*Myrtus luma*, *M. ugni* and *M. chequen*) and the currently overlooked conifers, Prince Albert's yew (*Saxegothaea conspicua*), *Pilgerodendron uviferum* and the alerce (*Fitzroya cupressoides*). All these new finds were grown in the Exeter Nursery before being sold to eager gardeners. Many of these introductions thrived in the mild Cornish climate, where mature specimens still make a lovely splash of colour in spring and summer. Elsewhere, their slightly tender nature requires a cool conservatory.

Meanwhile Thomas Lobb, inspired by his brother's adventures, determined also to become a plant hunter and approached his employer. The timing was perfect, because James junior was eager to gather a collection of tropical plants, especially orchids. It was his aim to exploit the expanding market for tender plants and to experiment with hybridizing. Once again Sir William Hooker was consulted, and all were in agreement that the East Indies offered rich pickings. Therefore Thomas, aged thirty-two, was dispatched from Portsmouth in January 1843 on what turned out to be a four-year expedition. He had to stop at Singapore for the visa required to enter Java, and while waiting he explored Singapore, Penang, and Mt Ophir (now Gunung Ledang) on the Malay Peninsula.

Once again the lack of records makes it impossible to follow the journey in detail, but one can picture an expedition into the tropical rainforests. Thomas

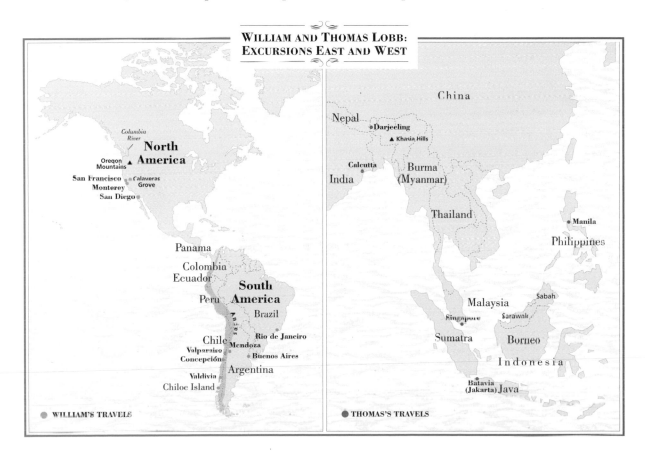

WILLIAM AND THOMAS LOBB:
EXCURSIONS EAST AND WEST

Thomas Lobb spent many months hacking his way through the lush vegetation of tropical rainforests, like this one in Sarawak, searching out orchids and pitcher plants in the green twilight of the jungle undercanopy.

assembles his team at the foot of Gunung Ledang, which he hopes to climb. Crossing the rough ladangs or clearings made by the natives to grow crops, they enter the forest of immense trees, which soar 130 feet or more to the unbroken canopy above. The undergrowth is a sparse mixture of palms, gingers and shade-tolerant plants. The party stops to shelter from the thunderstorms that sweep in after midday, before continuing along winding muddy paths. As well as raising the humidity to almost unbearable levels in the still air, the rain has brought out the leeches. These bloodthirsty creatures attach themselves to every nook and cranny and make life very uncomfortable. After a day's hard march, camp is made beneath a makeshift shelter of poles and palm leaves and Thomas is left with his thoughts as the guides disappear to hunt wild boar or deer for supper. Night falls after the briefest of twilights, silencing the cicadas only for the night shift of crickets and tree frogs to take over – the jungle is never quiet.

Echoing calls of gibbons greet first light, and looking up Thomas can make out orchids, *Nepenthes* species and ferns growing tantalizingly out of reach. The party continues trekking through the jungle, gradually gaining altitude. Eventually they hack through brakes of bamboo and enter the cloud forest. Here the wraith-like swirls of mists conceal smaller trees smothered in hanging moss and lichen, the branches draped in a rich array of epiphytic orchids, including a *Coelogyne* species with airy sprays of white flowers. Luxuriant ferns grow everywhere, and higher still are beautiful rhododendrons, their spent flowers carpeting the jungle floor. Scrambling among shrubs is the insectivorous *Nepenthes sanguinea*, looking prehistoric with its blood-red pitchers.

Above the cloud forest on the exposed mountain ridge, Thomas enjoys a view over unbroken rainforest. Close by, the golden-yellow flowers of the terrestrial orchid *Spathiglottis aurea* shine above the dwarf conifers (*Dacrydium* species) and tea trees (*Leptospermum* species). As he reaches the summit (4,186ft), he passes through bushes of *Rhododendron jasminiflorum*, covered with a mass of pretty white flowers. By mid-afternoon clouds have shrouded the mountain in cold mists and persistent drizzle. After an uncomfortable overnight stay, soaked to the skin and shivering, Thomas turns back into the steamy forest below to gather the lowland species he noted on the ascent.

These were Thomas's first orchids, pitcher plants and rhododendrons. Java and Burma were to prove equally rich hunting grounds. From Java came *Vanda tricolor* var. *suavis*, *Bulbophyllum lobbii*, the enormous *B. becari* (which smells revoltingly of putrid fish), *Phalaenopsis amabilis* and *Rhododendron javanicum*, while Burma yielded dendrobiums, *Paphiopedilum villosum* and *Rhododendron veitchianum*, which along with *R. jasminiflorum* was extensively used in later hybridization.

Thomas returned home in late 1847. After a year spent settling his collections in the nursery and seeing his brother for the first time in eight years, he set off east again on Christmas Day 1848. He landed in Calcutta in March 1849 but unfortunately missed the opportunity to take up an kind offer made by Sir Joseph Hooker. In those days plant hunters were very protective of their 'territories', but Sir Joseph, back in Darjeeling after his first trip into adventure, had written to his father in early 1849:

Tell Veitch by all means to send Lobb to Darjeeling before October if possible, he shall have every opportunity, facility and information I can afford, both as to living

Conservatories became an almost obligatory feature of larger Victorian gardens.
The most impressive example of all was the Crystal Palace, the extravagant centrepiece
of the Great Exhibition of 1851.

and collecting. May use my collection as much as he pleases in instructing himself on his own. I hope to return to these parts in October for seeds, and I will … let Lobb accompany me, when he shall be shewn everything I can shew, and have every facility I can afford subject to whatever advantage to Kew you may think fair – it is a chance Lobb may never get again, certainly never so cheaply – You must tell him I travel as a poor man and Lobb must not expect great tents and serv'ts.

Unfortunately Thomas did not receive this letter and carried on to Sarawak, the Philippines and Burma, where he found many more pitcher plants, a plethora of orchids, two medinillas and several more tropical rhododendrons. Somewhere along the line he and the letter met up, and on his return journey in March 1850 he stopped off in Calcutta. He travelled to Darjeeling, only to find Hooker had already left. Eventually the two men did meet in the Khasia Hills, with Hooker dryly commenting in his *Journal*: 'Yesterday, Thomas passed me with his "circus".' It would seem that Hooker's travelling companion, Dr Thompson, did not have a favourable opinion of Thomas, describing him as 'Very modest and well-behaved in his deportment but dreadfully conceited. He pooh-poohed Sikkim and had a very poor opinion of Lindley and Wallich!!!' Thomas, a bucolic Cornishman who was described by James Veitch as 'modest and retiring, of few words', may simply have felt ill at ease in such distinguished company.

Many plants of the beautiful *Vanda caerulea* that Thomas Lobb collected died in transit, but those that survived the journey fetched the enormous sum of £300. This kind of expenditure reveals how fashionable tender orchids had become. The mania for orchids grew to such proportions that in 1910 the Hon. Alicia Amherst noted that the natural habitat of many of them had been destroyed to feed the public demand and that at auctions successful bidders would pay over 1,000 guineas for a tiny plant of a new variety. Tender and stove house (hot house) plants were housed in the conservatory (or on a larger scale, the winter garden), which during the second half of the nineteenth century had become an essential garden adornment. Although orangeries and hot houses had existed before, it can be argued that the Crystal Palace helped to popularize the conservatory. Many of the over 6 million visitors would have travelled home dreaming of creating their own mini-Crystal Palace, as new technology made this dream a practicable and affordable reality for the suburban gardener as well as for the wealthy landowner. The invention in 1832 of cylinder glass by Lucas Chance resulted in larger panes, and the repeal of the Glass Tax in 1845 caused an 80 per cent drop in glass prices by 1865. Cast iron was now used to make glazing bars, which allowed exotic new curvilinear structures (such as the Palm House at Kew and the Great Conservatory at Chatsworth) and the manufacture of boilers and pipes for circulating hot water (and later steam) heating systems. Now, for the first time, the temperature within the conservatory and greenhouse could be accurately controlled. This meant that with sufficient wealth and enthusiasm one could build a range of greenhouses or conservatories, and in each create the perfect growing conditions for a specific group of plants, for example orchids, epiphytes, carnivorous plants, ferns and tender rhododendrons. These heating systems were also used to warm the house, which in turn allowed tender plants to be grown indoors for decoration.

Back in India, Thomas gathered several pleione orchids and *Hypericum hookeri* in the Khasia Hills, and after a period in the Nepalese foothills, where he found *Cardiocrinum giganteum*, he set sail back to Britain. After a brief stop in south-west India, where he found more orchids in the Nilghiri Hills, he arrived back in Cornwall in 1853.

While Thomas had been exploring the tropical Far East, William had made a third trip to the American continent (1849–53). This time he sailed to California, primarily to collect large quantities of seed from Douglas's conifers. It was to be an eventful trip, with things getting off to a bizarre start. He arrived to find San Francisco harbour choked with 200 abandoned ships. This was 1849, Gold Rush fever was endemic, and the crews had all jumped ship to seek their fortune along with the thousands of frenzied 'forty-niners' pouring in from every part of the country. The town was lawless and William quickly left in search of 'horticultural gold'.

Moving south to San Diego, he found the striking Santa Lucia or Silver fir (*Abies bracteata*) and *Rhododendron occidentale*, a deciduous species with lovely autumn foliage and scented, funnel-shaped, cream to pale pink flowers, a parent of many beautiful hybrids. He continued south-east, and from his base at Monterey in California he explored the Santa Lucia Mountains in 1850 and the San Juan river region in 1851. In the autumn he moved north, collecting large quantities of seed from Douglas's Monterey pine (*Pinus radiata*), sugar pine (*Pinus lambertiana*) and white western pine (*Pinus monticola*). He also gathered seed from the knobcone pine (*Pinus attenuata*) and what is now the world's tallest tree, the coastal or Californian redwood (*Sequoia sempervirens*). This had been discovered by Archibald Menzies in 1794 but was first introduced into Russia in 1840, three years before seed was sent to Britain. It is a long-lived tree – a specimen felled in 1934 was estimated to be 2,200 years old. At present the world's tallest tree is reputed to be the 'National Geographic Society tree', which in 1995 measured 365 feet 6 inches.

The following year William moved still further north into Douglas's stamping ground of the Oregon Mountains and the Columbia river, where he gathered seed of the noble fir (*Abies procera*) and the Douglas fir (*Pseudotsuga menziesii*) and discovered three new conifers overlooked by Douglas: the graceful western red cedar (*Thuja plicata*), the Californian red fir (*Abies magnifica*) and the Colorado white fir (*Abies concolor*). On his way back to San Francisco he collected seed from Douglas's giant fir (*Abies grandis*) and western yellow pine (*Pinus ponderosa*). In northern California he found two more conifers – the Californian juniper (*Juniperus californica*) and the Pacific white fir (*Abies concolor* var. *lowiana*). Other introductions that make a splash of colour in the garden were the lovely yellow-flowered *Fremontodendron californicum*, the red *Delphinium cardinale* and the blue *Ceanothus* × *lobbianus*. As with the monkey puzzle, William was the first to gather seed in bulk from trees that were still rare in cultivation. The amount of viable seed he sent home meant that Veitch & Sons was able to grow literally thousands of seedling trees.

William's third trip, however, remains most notable for another giant. During the summer of 1852 he was back in San Francisco, probably sending his latest collection to Britain and enjoying some much earned rest. One evening he was a guest at the summer meeting of the newly founded Californian Academy of Science when he heard a story that sent a shiver of excitement down his spine. Dr Albert Kellogg, the Academy's founder and a keen amateur botanist, introduced a hunter named A. T. Dowd, who had brought to him a specimen of a new type of tree and a strange story. Dowd told the rapt audience that earlier that year he had been out chasing a large grizzly bear in the foothills of the Sierra Nevada in Calaveras County. The chase was long and hard and Dowd followed the bear into a strange new part of the forested hills. Suddenly, and to his amazement, he stumbled into a grove of gigantic trees. Losing interest in the bear, he wandered around in astonishment at the sheer size of the monsters in front of him. When he returned to camp he told his story to his companions, most of whom did not believe him and accused him of being drunk. A few less sceptical souls allowed themselves to be led to the grove, where they were equally astonished by the magnitude of the trees.

The impact that such a tree would have on the British gardening fraternity, and the importance that Veitch & Sons would attach to being the first nursery to offer it for sale, must have struck William at once. As soon as the meeting ended he hurried to the Sierran foothills. All that he had seen on his extensive travels did not prepare him for Calaveras Grove, but unfortunately his prose does not do the moment justice: 'From 80 to 90 trees exist all within circuit of a mile, from 250ft. to 320ft. in height, 10–20ft. in diameter.'

Excitedly William collected seed, botanical specimens, vegetative shoots and two seedlings. Cutting short his contract, he booked a passage on the first ship home, arriving back in England in December 1853. He knew he risked Veitch's displeasure by coming back prematurely, but he also knew that it would be worse if someone else got back with seed before him. In this he was correct, and a discerning eye for good garden plants was one of the characteristics that made both brothers such successful plant hunters. Veitch was ecstatic with William's new find, and by the summer of 1854 the nursery was offering seedlings at 2 guineas each or 12 guineas a dozen. The Victorians fell in love with the Wellingtonia, and it became as much of a fad as the monkey puzzle, using it as a specimen tree and often planting it to form avenues.

Veitch entrusted the classification and naming of the tree to John Lindley at the Horticultural Society, who named it *Wellingtonia gigantea* in honour of the recently deceased and much lamented Duke of Wellington. Not surprisingly, Dr Kellogg was furious, arguing it should be called *Washingtonia* after George Washington, war hero and first President. The quarrel raged for years before the less evocative but non-partisan name of *Sequoiadendron giganteum* was allocated to show the tree's botanical relationship with the coastal or

Calaveras Grove, where William Lobb discovered *Sequoiadendron giganteum,* became a major tourist attraction. This woodcut shows a group of sightseers dancing on the stump of a felled tree.

Californian redwood (*Sequoia sempervirens*). Its common name remains Wellingtonia or giant sequoia. Although William is credited with the introduction because he had the plant classified, he had been pipped at the post by John Matthew, from Perthshire in Scotland, who arrived back four months before William and distributed seed among his fellow lairds.

Calaveras Grove soon became a Californian tourist attraction, complete with hotel. Predictably several trees were felled to provide 'novelty' attractions, such as a dance floor on the cut stump and a trunk made into a bowling alley. The bark of one 116 foot tree was sent to England, where it was displayed in the Crystal Palace at its relocated site at Sydenham. Today Calaveras Grove is part of the National Park, which is famous for the tree called 'General Sherman'. Generally acknowledged to be the world's largest living organism with an approximate age of 3,200 years, 'General Sherman' is 275 feet tall with a girth of 82 feet at 4½ feet; it has a total trunk volume of 50,000 cubic feet and is estimated to weigh 2,500 tons.

William returned to California in the autumn of 1854. Friends and family pleaded with him not to go, for they saw that the years of hard travel had taken their toll and that he was not well. However, being the stubborn man he was, William brushed off their entreaties and took on another three-year contract. The fashion for conifers was in full spate, and much of his time was spent in California gathering seed from his and Douglas's conifers. After his contract expired he remained in California and nothing more was heard from him after 1860. He died at St Mary's Hospital and was buried in San Francisco on 5 May 1864. Perhaps it is most appropriate to leave the final word about William to his employers. As James Herbert Veitch succinctly put it in *Hortus Veitchii*: 'The singular success which rewarded his researches is, perhaps, unparalleled in the history of botanical discovery; the labours of David Douglas not even forming an exception.'

In August 1854 Thomas Lobb returned to Java. Here he secured some new Japanese plants, including the conifers *Cryptomeria japonica* 'Lobbii', *Cryptomeria japonica* 'Elegans' and the Japanese umbrella pine (*Sciadopitys verticillata*). These were acquired from the Dutch East India Company Botanic Garden at Buitenzorg (now Bogor), in a transaction negotiated with the Dutch authorities by Veitch before Thomas set out. Thomas now planned to explore Mt Kinabalu, Sabah in north Borneo, where Veitch had requested he search out the largest of the pitcher plants, *Nepenthes rajah*, reported to reach a circumference of nearly 19 inches, but his plans were disrupted by problems with the local people, and he cut short his trip, returning home in 1857. The pitcher plant was finally introduced in 1878 by two later Veitch plant hunters, Peter C.M. Veitch and F.W. Burbidge.

After another year at home, Thomas made what was to be his final expedition, returning to north Borneo, Burma, Sumatra and the Philippines in 1858 to collect foliage plants, which Veitch's sold as house plants. These were used to decorate window ledges or were placed in ornamental Wardian cases, which had become a favourite feature of the parlour. His finds from Borneo included the extraordinary and prize-winning *Alocasia lowii* var. *veitchii* and several davillias, and from Burma, the lovely *Lygodium polystachyum* and the first variegated fern, *Pteris argyraea*. According to James Veitch in *Hortus Veitchii*, Thomas lost a leg in 1860 after suffering exposure in the Philippines. Another, less romantic, account puts the amputation at a later date, with the operation conducted on his sister Jane's kitchen table in Cornwall.

Whatever the truth, Thomas retired to Devoran on the money made from his herbarium collections and through letting several cottages he had built. His last contact with Veitch's was in September 1869. The subject discussed is not known, but by the end of the night James junior was dead of a heart attack and Thomas returned to Cornwall for good to spend twenty-five quiet years in a secluded cottage, tending his garden and painting. He died peacefully and was buried in Devoran churchyard on 3 May 1894. Once again the last word should be left to James Herbert Veitch, who wrote that 'his efforts were not short of that of his brother'.

In terms of impact on the garden, the hardy species introduced by William became popular and many remain so today. In contrast, many of Thomas's tender species are grown only in botanic gardens and the hot houses of a few enthusiasts. This is not a fair picture. In their day Thomas's orchids, tender rhododendrons and foliage plants caused as much of a sensation as William's monkey puzzle, Wellingtonia and Chilean fire bush. It can even be argued that Thomas's plants were more important in horticultural and economic terms because they were the parents of many new hybrids created by Veitch's expert plant breeders. What is without doubt is that Veitch's decision to send out plant hunters paid huge dividends right from the start, with William's hardy introductions balanced nicely by Thomas's tender ones.

The Lobb Plant Introductions

The date beside each plant name is the date of its introduction into Britain.

WILLIAM LOBB

Tropaeolum speciosum (c. 1845)

Tropaeolum = (Gk) *tropaion*, trophy
speciosum = (Lat.) showy

Spectacular, fiery orange-red, 1in (2cm) wide flowers festoon bright green, five-lobed hairy foliage throughout the summer, followed by unusual turquoise seed clusters. The beautiful flame creeper is a deciduous scrambling climber, dying back to a root stool in winter. Lobb also introduced *T. lobbianum*, with yellow to red flowers. *Tropaeolum* is the botanical name of the nasturtium.

Occurs in Chile, growing among scrub, spreading 6½ft (2m) or more.

Embothrium coccineum (1846)

Embothrium = (Gk) *en*, in; *bothrion*, a little pit (in allusion to the anthers being in pits of the calyx)
coccineum = (Lat.) scarlet

With abundant honeysuckle-shaped, brilliant scarlet flowers during May and early June, the Chilean fire bush is a glorious, variable, semi-evergreen, erect slender shrub or small tree with long 4–5in (10–17cm), light to grey-green leaves. The Lanceolatum Group, collected by Harold Comber, contains hardier, narrow-leafed, less evergreen forms.

Native to Chile and south-west Argentina, growing in open woodland at low altitudes and reaching from 32–48ft (10–15m). A yellow form occasionally occurs.

Berberis darwinii (1849)

Berberis = Latin form of the Arabic name for the fruit
darwinii = after Charles Darwin

With profuse, bright orange, cup-shaped flowers in drooping clusters in April and May, this is an evergreen shrub with ½–1in (1–2cm) holly-like pointed leaves, some of which colour bright red in autumn. Originally found by Charles Darwin on his *Beagle* voyage and introduced by Lobb fourteen years later. It is a parent to two good hybrids, B. × *stenophylla* × *B. empetrifolia*), with yellow flowers on graceful arching branches, and B. × *lologensis*, a superb natural hybrid with *B. lineariloia*, with apricot-yellow flowers.

Ceanothus × *veitchianus* (1853)

Ceanothus = Greek name for a spiny plant
veitchianus = after the Veitch nurserymen

Lovely deep blue flowers in neat rounded spikes envelop this fine, large, spreading evergreen shrub in May and June. It is a natural hybrid between *C. rigidus* and *C. griseus*, with small, oval glossy green leaves. Lobb also introduced *C.* × *lobbianus* (*C. dentatus* × *C. griseus*, 1853, bright blue flowers) and *C. papillosus* (1850, rich blue with longish leaves). The latter species has produced some good hybrids, e.g. 'Concha' × *C. impressus* (red buds, deep blue flowers), and 'Delight' × *C. rigidus* (long spikes of rich blue). All ceanothus are short-lived plants.

From Monterey, California, USA, growing in scrub near the coast, reaching 10ft (3m) tall.

Thuja plicata (1853)

Thuja = (Gk) *thuia*, a type of juniper
plicata = (Lat.) pleated

Shiny bright green foliage in flattened pendulous sprays clothes the western red cedar, a splendid large conical tree, with fissured, reddish bark shedding in strips. The leaves, which are never too dense and impart a lively billowing texture, colour bronze in winter and emit a delicious fruity fragrance when crushed. Low branches can touch the ground and take root to form a ring of new trees around the main bole.

Ranges from the south Alaskan coast of North America to north-west California and from British Columbia to northern Idaho, predominantly on rich, very moist (even waterlogged) soils, reaching 165ft (50m) tall.

Sequoiadendron giganteum (1853)

Sequoiadendron = *sequoia*, after the inventor of the Cherokee alphabet; *dendron*, (Gk) tree
giganteum = (Lat.) unusually large or tall

Majestic giant trunks with thick, reddish-brown, furrowed bark support graceful down-sweeping branches covered in spirals of small, bright green, aniseed-scented leaves. It is an exceptional species, forming a very tall, massive conical tree.

Ranges along the western Sierra Nevada, California, USA, in eighty preserved groves and is exceptionally long-lived, with authenticated ages of over 3,500 years and heights of 225–295ft (75–90m). No felling has occurred this century, because of its poor quality timber.

THOMAS LOBB

Phalaenopsis amabilis (1846)

Phalaenopsis = from (Gk) *phalaina*, moth; *opsis*, like
amabilis = (Lat.) lovely

Elegant 3ft (1m) long sprays of lovely 4in (10cm) wide, purest white flowers, the lips marked with a variable yellow stain and red spots, the whole looking like a giant white moth. This species of epiphytic moth orchid has beautifully marbled, reddish and green broad leathery leaves and is a parent to most modern hybrids. Multitudes of spectacular hybrids exist. Lobb also introduced the natural hybrid *P.* × *intermedia* (*P. aphrodite* × *P. equestris*, 1852) from the Philippines.

Widespread from Java through the Philippines to New Guinea and northern Australia, growing in rainforests.

Nepenthes sanguinea (c. 1847)

Nepenthes = from Greek literature, any plant that could produce euphoria, a reference to its medicinal properties
sanguinea = (Lat.) blood-red

Exquisite cylindrical pitchers, suffused and spotted blood red, have glossy deep red mouths. This desirable climbing perennial scrambles through shrubs or low trees, and the bright green thick leaves and stems are flushed red. Lobb also introduced *N. albo-marginata* from Borneo.

From the Malay Peninsula, growing in highland cloud forest and scrub above 3,200ft (1,000m) and spreading 10ft (3m) or more.

Aerides rosea (1850)

Aerides = from (Lat.) *aer*, air
rosea = (Lat.) rose-like

Fabulous pendant sprays of rose-pink flowers, overlain with fine purple spotting, 18–24in (45–60cm) long appear in late spring. The fox-brush orchid is a handsome epiphyte, with 4–10in (10–25cm) stems and bright green 6–14in (15–35cm) linear, leathery leaves. Lobb also introduced *A. multiflora* from India.

Ranges from north-east India to northern Vietnam and southern China, growing in forests and often even in disturbed habitats.

Rhododendron veitchianum (c. 1850)

Rhododendron = (Gk) *rhodo*, rose; *dendron*, tree
veitchianum = after the Veitch family of nurserymen

Beautiful, 5in (12cm) wide, white, yellow-throated crinkle-edged fragrant flowers appear from February to July. A very fine small evergreen, usually growing as an epiphyte. A popular Victorian houseplant. Thomas Lobb introduced several other tender species, e.g. *R. malayanum* (*c.* 1850, small pink flowers), *R. brookeanum* (*c.* 1850, golden-yellow flowers) and *R. javanicum* (*c.* 1850).

Ranges from southern and central Burma to Thailand and Indo-China, growing in mixed oak and dry evergreen forests at 2,950–7,800ft (900–2,400m), reaching only 3ft (1m) tall.

Vanda tricolor var. *suavis*, one of the lovely orchids sent back by Thomas Lobb from the lush rainforests of Java.

Vanda caerulea (1850)
Vanda = from the Sanskrit for 'epiphytic orchid'
caerulea = (Lat.) dark blue

Glorious 4in (10cm) blue flowers, marked with deeper blue tessellations appear on branched flowering stems in winter. This is a most coveted large, epiphytic species, with 10in (25cm) long light green leaves. It has been used extensively in breeding, e.g. *V.* 'Rothschildiana'. Lobb also introduced the lovely *V. tricolor* (1846) from Java, which has fragrant pink-lipped pale yellow, brown-patterned flowers.

Ranges from India to Burma and Thailand, growing in upland rainforests and reaching from 27in– 4½ft (0.75–1.5m). Formerly common, wild populations have been largely destroyed through over-collection.

7

CHINESE PUZZLE

Ernest Wilson

(1876–1930)

IN APRIL 1899 AN UNTRAVELLED TWENTY-THREE-YEAR OLD WAS dispatched to China with instructions to find and collect a beautiful tree. The plant hunter was Ernest Henry Wilson, his employer was James Veitch & Son, and the sought-after specimen was the handkerchief or dove tree (*Davidia involucrata*). On previous occasions the Veitch empire had been innovative in its strategy of sending out collectors, but this time Sir Harry Veitch was reacting to circumstances rather than dictating them. By the late 1890s the diversity of China's flora was becoming increasingly apparent, largely because of the work of three amateur botanists who had been sending back dried herbarium specimens from China's vast, unexplored interior.

Left: Wilson introduced thousands of garden plants, including the climber *Clematis montana* var. *rubens.*

Above: Ernest Wilson: a self-portrait of the most prolific of all the plant hunters.

Improved access into China had been grudgingly granted in 1860 following the wars with Britain and was quickly acted on. Many of the first travellers were French missionaries, two of whom sent dried plant collections to the Musée d'Histoire Naturelle in Paris. Abbé Jean Pierre Armand David (1826–1900) arrived in 1862, but it was on a trip to Szechwan (now Sichuan) and Mupin (now Baoxing) in the Tibetan borderland (1868–70) that he found the tree that bears his name. He was also the first westerner to observe the giant panda in the wild, and he arranged for the capture of an adult bear and its transportion to Paris (where, sad to say, it soon died).

David returned to France in 1874 with an impressive 250 new plant species and ten new genera found, but not necessarily introduced. His work was continued by a fellow missionary, Jean Marie Delavay (1838–95), who spent ten years in north-east Yunnan and botanized the area thoroughly. The third amateur collector was a Scot, Augustine Henry (1857–1930), who, as an assistant medical officer of the Imperial Chinese Maritime Customs Service, took up collecting plants in 1882 in order to relieve the boredom of his posting at Ichang (now Yichang). Henry sent his herbarium specimens to Kew, and by the time of his return in 1899 he had collected 158,000 specimens, which equated to over 500 new species, twenty-five new genera and even one new family, Trapellaceae. This plethora of new material came to Sir Harry's attention, and it was Henry's offer of help to any plant hunter sent out that finally persuaded Veitch to send another collector to China.

Ernest Wilson was born on 15 February 1876 and grew up in the picturesque Cotswold village of Chipping Campden. At sixteen he began an apprenticeship at Hewitt's nursery in Solihull, then moved on to the Birmingham Botanic Garden at Edgbaston. While in Birmingham he took classes at Birmingham Technical College to improve his botanical knowledge. He won the Queen's Prize for Botany which helped him to progress to the Diploma course at Kew. Wilson intended to become a botany teacher, but the Director of Kew, W. T. Thiselton-Dyer, when asked by Sir Harry to recommend a suitable man to go to China, put forward Wilson's name.

Before leaving Britain, Wilson spent six months increasing his horticultural knowledge at the Veitch nursery at Coombe Wood. Surrounded by the evidence of the Veitch legacy of plant hunting, Wilson must have felt some trepidation, but he shared many of the characteristics of his predecessors. He was practical, hard-working and well organized, had a good eye for a garden plant, was a natural diplomat and was wise enough to take advice when offered.

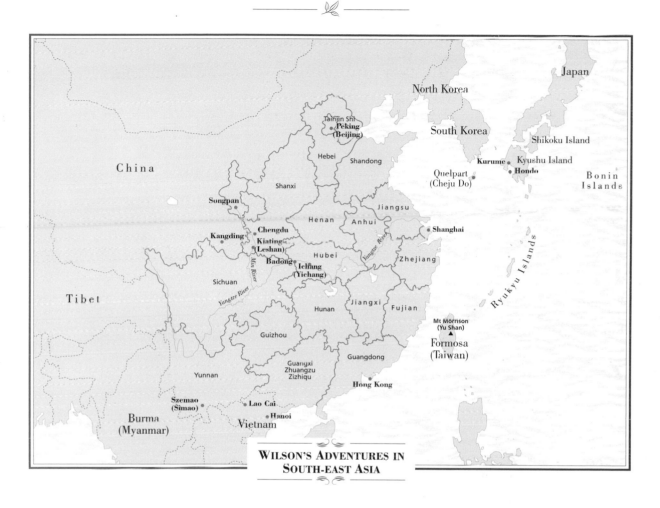

WILSON'S ADVENTURES IN SOUTH-EAST ASIA

He left Liverpool aboard the *Pavonia*, bound for America, and broke his journey at the Arnold Arboretum, where he met Charles Sargent, who was to play a large role in his later career. Continuing his journey across the continent by train, Wilson sailed from San Francisco for China on 6 May. He was eager to reach the small outpost of Szemao (now Simao) in Yunnan province, as he knew Augustine Henry was about to return home. When Wilson arrived in Hong Kong on 3 June, however, he found himself caught up in the midst of an outbreak of bubonic plague; no Chinese were allowed to leave the colony, so he was forced to leave for Hanoi in French Indochina without an interpreter. In Hanoi he met an English-speaking Frenchman who warned him of trouble brewing in the region, and when Wilson sailed up the Red River to Lao Cai he found that the situation had indeed deteriorated. He was forced to wait several dull weeks there, enduring sweltering heat and the constant threat of malaria.

At the end of the 1,000-mile journey from Hong Kong, the exhausted Wilson was welcomed by Henry. He was somewhat taken aback at Henry's help, which amounted to a scruffy piece of paper showing a crude map. On this map, which was of such a small scale that it covered an area of 20,000 square miles, was roughly marked the position of a single *Davidia involucrata*. Undaunted, Wilson planned an expedition to Sichuan to find the elusive tree. Arriving at Ichang on the Yangtze on 24 February 1900, he began to assemble his team. For such a venture a large entourage was unavoidable – Wilson needed interpreters, guides and porters to carry his collecting equipment, scientific instruments and food and medical supplies. He travelled with two sedan chairs, one for himself and one for his headman, but these were rarely used.

To make the journey upriver easier, Wilson bought a Chinese houseboat and set off on 15 April 1900 for Badong. Here he heard that anti-European feelings were strong, but he was confident that his destination was sufficiently remote to avoid the worst of these problems. From Badong he set off into the hills to find the X on Henry's map. When he arrived at the spot, he gazed in horror at a smart, new wooden house standing next to the stump of a *Davidia involucrata*. Wilson was mortified – he had travelled 13,000 miles for nothing and in despair he wrote, 'I did not sleep during the night of April 25th 1900.'

Wilson retreated to Ichang, where he tried to console himself by exploring the nearby hills. Even the discovery of the curious edible climber *Actinidia chinensis* (now known erroneously as the kiwi fruit) did not dispel his depression. Then (as was often the case with Wilson) good luck smiled upon him. Clambering through dense woods on 19 May he suddenly came across a magnificent handkerchief tree in full flower. While this must have made his spirits soar, Wilson's jubilation is hard to detect in his mundane prose: 'To my mind *Davidia involucrata* is at once the most interesting and beautiful of all trees of the north-temperate flora … The flowers and their attendant bracts are pendulous on fairly long stalks, and when stirred by the slightest breeze they resemble huge Butterflies hovering amongst the trees.' Wilson gathered a large quantity of the nutmeg-shaped seeds and with a light heart turned his attention to the new plants that were all around. He found literally hundreds of trees, shrubs and herbaceous species, including *Acer griseum*, (the paperbark maple), *A. oliverianum*, *Abies fargesii*, *Betula albo-sinensis*, *Lonicera tragophylla*, *Viburnum rhytidophyllum*, *V. utile*, *Clematis armandii*, *C. montana* var. *rubens*, *Magnolia delavayi*, *Rhododendron decorum*, *R. fargesii*, *Camellia cuspidata* and *Rodgersia aesculifolia*.

The beautiful white bracts of the handkerchief or dove tree, *Davidia involucrata*, look ghostly set against fresh green spring leaves.

With his collections carefully packed and dispatched, Wilson set sail for England, arriving home in April 1902. Sir Harry Veitch was delighted with yet another selection of novelties to sell at premium prices, and he presented Wilson with a gold watch to commemorate his achievements. Veitch's celebrations were premature, however, for it transpired that a Frenchman, Paul Guillaume Farges, had in fact returned to Paris in 1897 with seed of *Davidia involucrata*, one of which had germinated. Denied of their 'introduced by' epithet, worse was to follow for the Veitch Nursery when Wilson's seeds apparently failed to germinate. Unknown to Veitch, the seed requires a period of up to eighteen months of alternating cold and warmth to stimulate germination. At last, to their great relief, the seedlings were finally seen to be coming up and thousands of plants were successfully raised.

Wilson married his sweetheart, Helen (Nellie) Ganderton, in June 1902, but there was to be no extended honeymoon for the newly-weds. Wilson was contracted by Veitch to undertake a second trip to China, and on 23 January 1903 he left Britain on a two-year journey. The aim this time was to track down the alpine yellow poppywort (*Meconopsis integrifolia*), which grew in the mountains of Tibet. In Shanghai Wilson again recruited a large team of helpers:

> I engaged about a dozen peasants from near Ichang. These men remained with me and rendered faithful service during the whole of my peregrinations. Once they grasped what was wanted they could be relied upon to do their part, thereby adding much to the pleasure and profit of my many journeys. When we finally parted it was with genuine regret on both sides. Faithful, intelligent, reliable, cheerful under adverse circumstances, and always willing to give their best, no men could have rendered better service.

Wilson bought another houseboat, the *Ellena* (named after his wife), and sailed up the Yangtze to Ichang. Travelling further than on his previous trip, he entered a succession of magnificent gorges full of flowering shrubs. 'The scenery hereabouts is savagely grand and awe-inspiring,' he noted, but he had little time to enjoy the spectacle because the treacherous Yangtze rapids were fast approaching. The party was forced to shelter from fearsome storms before dragging the boat over the lower rapids. When they reached the Yeh-Tan rapids, Wilson watched in horror as three of the boats ahead of him were dashed to pieces on the jagged rocks and the crews drowned. His crew frantically tried to appease the water gods: 'My captain chin-chinned, joss crackers were exploded, a little wine and rice were thrown over the bow, joss sticks were burnt, together with candles and some paper cash – in short, every rite necessary to appease the terrible water-dragon was strictly observed.'

The appeals worked, and with the combined efforts of 100 men pulling on bamboo ropes the boat was hauled into the calm waters beyond the cataract. Wilson realized just how fortunate he had been when, looking back, he saw the next boat capsize and watched helplessly as two men drowned. However, a few days later his luck ran out when one of his own men was drowned following a collision with a small boat. As a result of the collision, the *Ellena* slipped her mooring and almost came to grief in a whirlpool, but with a great effort Wilson's men managed to drag the boat to safety. Wilson raced after the small boat and angrily confronted the owner – the man was reported to the local magistrate and fined for his carelessness rather than for the loss of a man's life!

Reaching Kiating (now Leshan) in mid-June, Wilson explored the rain-soaked Wa-Shan Mountains and gathered 200 species before heading due west to Hanyuan, set in its deep, humid valley. He had not lost sight of his objective, however, and set off to track down the yellow poppywort in the mountains near Kangding on the Sino-Tibet border. On 14 July 1903, after three weeks of tough travel, he arrived in the dirty town of Kangding and, after two days set off along the Lassa (now Lhasa) road. Passing lamaseries and fields of ripening crops, he crossed the Ya-chia Pass (over 9,840 feet). Wilson began to suffer from altitude sickness, and to make matters worse he was forced to spend a very uncomfortable and stormy night in a leaking hut. At 2a.m. he awoke, minus his blanket: 'pulling it over me again I disturbed four half-drowned chickens whom my thoughtless men had tied to a post alongside my bed. These chickens, resenting the loss of the blanket, tried to follow it and succeeded in nearly blinding me with mud.'

The following morning, warmed by the bright sun, Wilson climbed the surrounding hills. Soon he came across a dazzling sheet of yellow *Meconopsis integrifolia*, their papery blooms fluttering in the mountain breeze. Once more his matter-of-fact prose records the event with masterly understatement: 'From 11,500 feet to 13,000 feet the gorgeous *M. integrifolia*, growing 3 feet tall, with its peony-like, clear yellow flowers 8 to 11 inches across.' Buoyed up by his success, Wilson decided to seek out another species, the red poppywort (*M. punicea*). Leaving Kangding on 23 July, he headed for Songpan via the Min

Overleaf: A misty valley-side in Sichuan province, China.
The rich vegetation gave Ernest Wilson ample scope for plant hunting.

valley. It was somewhere in this dry valley that he first found the regal lily (*Lilium regale*), which caused such a sensation on its introduction that it was the subject of a future expedition (on which his luck finally ran out). Arriving at Songpan on 27 August, Wilson learned from the locals that his prize might be found in the Kung-lung Pass, and with five mounted soldiers he travelled nineteen miles into the mountains, where he found an abundance of delicate red poppies on the exposed hillsides above 11,500ft. By now Wilson had travelled some 650 miles in ten weeks, and the physical exertions were beginning to tell. He had lost in excess of 3 stone and was suffering from exhaustion. But he was in positive mood when he returned to spend the winter in Ichang. Subsequent visits to Kangding in May and June 1904 yielded *Meconopsis henrici*, *Cypripedium guttatum*, *C. tibeticum* and many alpines, and a final foray near Leshan in November brought him the lovely *Dipelta floribunda*.

In March 1905 Wilson arrived back in Britain, bringing with him seed of 510 species, including *Primula pulverulenta*, *Viburnum davidii*, *Rhododendron calophytum*, *R. lutescens*, *Rosa moyesii* and 2,400 herbarium specimens. For his efforts this time, Sir Harry presented Wilson with a gold pin, shaped like the flower of a *Meconopsis* species and encrusted with forty-one diamonds. During the winter of 1905–6 Wilson was visited by Charles Sargent, who was still very keen that the Arnold Arboretum should send its first plant hunter to China. Wilson was not interested, preferring to stay at home with his wife, who was pregnant, but Sargent used all his powers of persuasion. Eventually Wilson capitulated, but only after he had secured good terms – a salary of £750 a year for two years and the possibility of work at the Arnold Arboretum on his return. On 21 May 1906 Helen gave birth to a daughter, christened Muriel Primrose, although Wilson did not have long to enjoy fatherhood, for by the autumn he was making his preparations, hoping to be in Ichang the following April. Sensibly, Sargent had the foresight to insist that he took a camera, the additional cost of a porter to carry it being 'not a very important item'. Wilson clearly had a talent for photography and the pictures taken of the people and landscape with his large whole-plate Sanderson remain very evocative.

On this, his third expedition, Wilson was returning to the scientific roots of plant hunting. His instruction from Sargent was to 'increase the knowledge of the woody plants of the [Chinese] Empire and to introduce into cultivation as many of them as is practicable'. In this way he was following the example of Banks and Hooker, who collected for the furtherance of botany and who, in the course of their work, stumbled across some beautiful garden plants. Once more

Above: One of Wilson's
photographs, showing the
man-power required to
collect plants in a distant
land. On the right is the
sedan chair that he used to
display his status, important
for foreigners in China.

Right: A relaxed-looking
Wilson and his team prepare
to head upriver aboard the
Harvard. By using a boat
he was able to transport his
plant collections and guides
efficiently and with a degree
of comfort.

Wilson crossed the Atlantic and travelled across the States by train. This journey was almost more hazardous than those he had experienced in China – he wrote to Sargent on 7 January 1907, just before leaving America:

> Just outside Omaha a switch engine ran into us derailing and smashing up the tender of our engine together with the luggage and dining cars … after a considerable delay … we proceeded on our way. Later the same day we passed two East bound express trains in collision. The day following we passed two wrecked freight trains. Altogether I can assure you that I have had enough of American railways to last me for a time.

The next day Wilson sailed from San Francisco aboard the *Doric*, arriving in Shanghai on 4 February. He headed inland and reached Ichang on the 26th. For a third time he purchased a river boat, naming her the *Harvard*, and set about assembling his collecting team. Many of those chosen had accompanied him on previous trips, and sought him out to offer their services. Unfortunately, this trip did not prove to be as enjoyable as his previous expeditions.

Initially all went well. Wilson spent the summer collecting in the Lushan Hills of Jiangxi province and found the impressive lily-like *Cardiocrinum cathayanum*. But in September he was struck down with malaria, and after spending the winter recovering, repairs to the *Harvard* prevented him from making an early start in the spring. Eventually, in early summer, he reached the prosperous city of Chengdu, in western Sichuan. The city was full of skilled craftsmen and artisans producing luxury goods, and Wilson marvelled at the temples and courtyard gardens full of trained plants, including, 'Two magnificent specimens of Crepe Myrtle (*Lagerstroemia indica*), trained into the shape of a fan some 25 feet high by 12 feet wide, and said to be over 200 years old, are finer than anything of the kind I have seen elsewhere.' After exploring Guan Xian in the north-west and Kangding in the south-west, that winter Wilson dispatched his first season's finds from Ichang. The plants arrived safely, bar one mishap. Wilson had cut corners in packing a consignment of 18,237 lily bulbs (to save money he had not coated each one in clay) and over 95 per cent of the bulbs rotted during transit.

Several new conifers were added to Wilson's collection in 1907, but foul weather and a poor seed set forced him to return in 1910 to gather another batch of seed. November brought news of the deaths of the Chinese Emperor and his wife, both in suspicious circumstances, and Wilson was worried that the country might yet again descend into anarchy. Much to his chagrin, news also reached him that Charles Sargent and his former employer, Sir Harry Veitch, had joined forces to send another plant hunter, William Purdom, to China. Wilson was not angry about Purdom's appointment so much as resentful that he had not been consulted. The two men met in Peking (now Beijing) in April 1909, when Wilson magnanimously shared his knowledge of plant hunting in China with Purdom, and on 25 April he dispatched his camera negatives to London and his impressive plant collection to Boston. His haul included *Acer wilsonii*, *Clematis tangutica* var. *obtusiuscula*, a hardy plumbago (*Ceratostigma willmotianum*), a dogwood (*Cornus kousa* var. *chinensis*), *Magnolia wilsonii*, *M. sinensis*, *M. dawsoniana*, *Picea likiangensis* and *Rhododendron mouipinense*.

Wilson travelled back to Europe on the Trans-Siberian railway, stopping *en route* to visit nurseries in St Petersburg, Berlin and Paris on behalf of the

A member of the dogwood family, *Cornus kousa* var. *chinensis* looks like a fresh fall of snow when in full flower.

Arnold Arboretum. Sargent kept his word, offering Wilson a temporary post to supervise the organizing of his herbarium collection, and on 21 September 1909 Wilson and his family left England for Boston. To his surprise, Wilson found that he was a celebrity in Boston society and in great demand. It was the Bostonians who gave him his nickname 'Chinese' Wilson, of which he was modestly proud.

The persuasive Sargent was not about to let his prize collector slip into dusty obscurity in the Arnold Arboretum, however, and he talked Wilson into a fourth trip to China, this time to re-gather conifer seed and collect more bulbs of the splendid regal lily. History does not recount what Helen thought of the idea of being left alone once more, this time in a new country; however, a compromise was reached and the family returned to England in February 1910.

Once again Wilson took the overland route via Russia, arriving in Peking in May and reaching Ichang (now Yichang) on 1 June 1910. By now a seasoned explorer, he did not take long to organize his trip. With some of his most experienced men, he left three days later to trek to the remote region of north-west Hubei, passing through woodlands of oak, pine and flowering shrubs and glens full of Henry's Boston ivy (*Parthenocissus henryana*). Near the small, isolated villages Wilson was astonished by the precipitous man-made cultivation terraces that crept up the mountainsides. In contrast with this rural idyll, the town of Xiantang was wretched, with overflowing open sewers (Xiantang ironically means 'fragrant rapid'). The party left as quickly as possible and a few days later stopped overnight at Peh-yang-tsai, where Wilson noted that 'the country people everywhere in these parts, were extremely nice and obliging, and it was a real pleasure to be amongst them'. The following morning he decided to climb the limestone peak of Wan-taio Shan, which towered above the village. The steep walk up through the forest of Chinese beech and *Davidia involucrata* was worth the effort, for on the way he collected forty species, including *Syringa julianae* at the summit. Back on the path once more, he found the peculiar 'ear-shaped and gelatinous' Jew's ear fungus. These were cultivated by the local villagers on oak logs, and although considered a delicacy, 'I tried them, but did not find them very palatable, and the experiment resulted in a bad stomach-ache!'

By early June Wilson had reached the hamlet of Li-erh-kou. A steep ascent brought him out on to a ridge 'where *Viburnum rhytidophyllum* luxuriates … with its long, thick wrinkled leaves'. For three weeks he toiled over moorland and struggled through rhododendron and bamboo forest before reaching the

Great Salt Road. Although the road was more defined, the going was punishing and the heat intense. At the tiny hamlet of Hsao-pingtze Wilson paused, breathless and tired. He was surrounded by 'range upon range of bare, treeless, sharp-edged ridges, averaging 5,000 to 6,000 feet in height, with outstanding higher peaks and grander ranges in the beyond … Never have I looked upon a wilder, more savage and less inviting region … It was indeed sufficient to awe and terrorise one. Such scenes sink deep into the memory and the impressive stillness produces an effect which is felt for long years to come.'

In early July Wilson reached the sizeable town of Xuanhan, and after a rest he decided to head westwards to Paoning (now Langzhong). Along the route the villages became more frequent, and he attracted considerable attention. Several times while shopping in the market he was treated as some sort of freak sideshow – huge crowds would gather to stare for hours, expressionless, with unashamed curiosity. On 20 August 1910, after a strenuous hike:

> we sighted the city of Sungpan nestling in a narrow smiling valley, surrounded on all sides by fields of golden grain, with the infant Min, a clear, limpid stream, winding its way through in a series of graceful curves. In the fields the harvesters were busy, men, women, and children, mostly tribesfolk, in quaint costume, all pictures of rude health, laughing and singing at their work. Under a clear Tibetan-blue sky, the whole country bathed in warm sun-light … gladdened the hearts of all of us.

Entering the arid Min valley, where 'in summer the heat is terrific, in winter the cold is intense, and at all seasons these valleys are subject to sudden and violent windstorms against which neither man nor beast can make headway', Wilson saw giant Chinese characters carved in the rocks warning of the danger of landslides. The narrow track was the main supply route to and from Songpan and mule trains toiled along it in the heat. But it was here that the:

> Regal Lily has her home. There in June, by the way side, in rock-crevice by the torrent's edge and high up on the mountainside and precipice this Lily in full bloom greets the weary wayfarer. Not in twos and threes but in hundreds, in thousands, aye, in tens of thousands. Its slender stems, each from 2 to 4 feet tall, flexible and tense as steel, overtop the coarse grasses and scrub and are crowned with one to several large funnel-shaped flowers, each more or less wine-colored without, pure white and lustrous on the face, clear canary yellow within the tube and each stamen filament tipped with a golden anther. The air in the cool of the morning and in the evening is laden with delicious perfume exhaled from every blossom. For a brief season this Lily transforms a lonely, semi-desert region into a veritable fairyland.

The regal lily is worthy of its title. The delightfully scented and elegant *Lilium regale* was one of Wilson's favourite introductions and remains popular today.

After a week's march Wilson established his camp and set about marking the position of over 6,000 bulbs which were to be lifted in October. When they broke camp on 3 September, Wilson, fatigued but happy, chose to celebrate his success by riding in his sedan chair along the narrow trail:

> Song was in our hearts, when I noticed my dog suddenly cease wagging his tail, cringe and rush forward and a small piece of rock hit the path and rebounded into the river some 300 feet below us. I shouted an order and the bearers put down the chair. The two front bearers ran forward and I essayed to follow suit. Just as I cleared the chair handles a large boulder crashed into the body of the chair and down to the river it was hurled. I ran, instinctively ducked as something whisked over my head and my sun hat blew off. Again I ran, a few yards more and I would be under the lea of some hard rocks. Then feeling as if a hot wire passed through my leg, I was bowled over, tried to jump up, found my right leg was useless, so crawled forward to the shelter of the cliff, where the two scared chair-bearers were huddled.

When he looked down he saw his leg was broken in two places and the flesh was torn and bleeding. The rock had also ripped off the end of his boot and his big toenail with it. Despite the intense pain he remained conscious and quickly instructed his men to make a splint from his camera tripod. As he lay helpless across the trail, a train of mules, unable to retrace the route, was forced to walk over the injured man. As Wilson sardonically noted, 'Then it was that I realized the size of the mule's hoof,' yet not one mule touched him.

Wilson was lifted into the second sedan chair and rushed by his worried team to Chengdu. By the time he arrived three days later, infection had set in and amputation looked likely. Once more his luck was in, for he was taken to Dr Davidson of the Friend's Mission, who, in an operation lasting 'more than an hour under chloroform', saved and set the leg. Wilson could not continue his explorations, however; indeed, it was three months before he could hobble along painfully on crutches. Nevertheless, his faithful team used their considerable experience to find some lovely species, including the Min fir (*Abies recurvata*), the flaky fir (*Abies squamata*), *Acer maximowiczii* and a graceful bamboo, which Wilson named *Sinarundinaria murielae* (now *Fargesia murielae*), after his beloved daughter. They also successfully lifted the lilies and packed them in clay for shipment.

Once Wilson was able to move, he dispatched 50,000 herbarium specimens and 1,285 packets of seed (Purdom managed only 304 in the same period of time) to Sargent. When identified, these added a further four new genera, 382 new species and 323 new varieties to the woody plant catalogue of China. Wilson was again offered work by Sargent when he arrived back in Britain in March 1911, but before taking up the post he underwent further surgery on his leg. It was re-broken and re-set, and although it healed well, he was left with a shortened right leg and a limp that made it necessary for him to wear an orthopaedic boot.

For two-and-a-half-years Wilson worked in Boston on his herbarium specimens and wrote the two-volume *A Naturalist in Western China*, published in 1913. Later that same year Sargent once again prevailed on him to make a plant hunting trip on behalf of Arnold Arboretum. Wilson's leg injury meant that an arduous expedition was out of the question, so Sargent suggested an extended trip to Japan, with an emphasis on conifers and cherries. Again the discussions between husband and wife are a matter of speculation, but clearly Helen could be as persuasive as Sargent, for on this trip Mrs Wilson became the first wife of a British plant hunter to accompany her husband on an expedition.

Ernest, Helen and Muriel arrived in Japan on 3 February 1914 and spent a year exploring the forests, taking photographs and investigating the Japanese nurseries. They were in the south during February and March, in central Japan between April and June, and moved to Hondo and Saghalin for July and August. The autumn was spent back in the central region, followed by a couple of months on the island of Shikoku. Wilson accumulated 2,000 herbarium species and took 600 photographs. By the time they arrived back in Boston in February 1915 the First World War had begun, but Wilson's injury (which he called his 'lily-limp') prevented him from enlisting. Further travel was delayed until 1917, and Wilson spent the intervening years working in Boston and writing two books, one on the cherries of Japan and the other on its conifers.

Wilson, with his family, set out on his sixth and last trip as a professional plant hunter in January 1917, heading once more for Japan. This time he explored the forests on Ryukyu Island in February and March, the Bonin Islands in April, and set off to Korea in May. For the rest of the year he explored the mainland peninsula and the islands of Quelpaert (now Cheju Do) and Warrior Island (now Ooryongto). This journey produced two maples (*Acer pseudosieboldianum* and *A. triflorum*), two syringas (*S. dilitata* and *S. velutina*), *Stewartia koreana*, *Spiraea trichocarpa*, *Astilbe koreana*, *Forsythia ovata*, the Korean arbor-vitae (*Thuja koraiensis*) and *Rhododendron weyrichii*. In January 1918 he moved on to Formosa (now Taiwan) and climbed the island's highest peak, Mt Morrison, where he found Eastern Asia's tallest tree, *Taiwania cryptomerioides*, and *Lilium philippinense* var. *formosanum*. In April he returned to Japan and visited the city of Kurume on Kyushu Island, where he saw a famous century-old collection of 250 named azaleas. When he was invited to the gardens of two leading specialists he found:

> veritable fairylands, and I gasped in astonishment when I realised that garden-lovers of America and Europe knew virtually nothing of this wealth of beauty. Most of the plants were trained into low standards, each about 20 inches high with flattened or convex crowns some 24 inches through, and were monuments to the patience and cultural skill of the Japanese gardener. The flowers, each about one-half to three-quarters of an inch across, were in such profusion as to almost hide the leaves. They are the roguish eyes of laughing, dimpled, and blushing blossoms.

Wilson had already encountered the Kurume azaleas in 1914 in a nursery just north of Tokyo, but this was their epicentre. After visiting what was claimed to be the original plant in an old garden, he climbed the sacred Mt Kirishima where he saw the Kyushu azalea (*Rhododendron kiusianum*) growing with *R. kaempferi* and a mass of their hybrids. These are considered the main parents of the Kurume hybrids, and his discovery had confirmed their origins.

These wonderful plants were a fitting climax to Wilson's plant hunting years and he wrote with feeling: 'Proud am I of being the fortunate one to introduce this exquisite damsel to the gardens of eastern North America.' Although in Britain the collection of Kurume azaleas became known as 'Wilson's Fifty', there were in fact, fifty-one varieties.

Wilson returned to the Arnold Arboretum in March 1919 a satisfied man. The Kurume azaleas were included in the crates of plants and seed, and he had discovered much about the flora of the previously poorly explored Korean peninsula. When he was offered the position of Assistant Director of the Arnold Arboretum, he and his family decided to settle permanently in Boston. He undertook a two-year promotional tour on behalf of the Arboretum and he wrote several more books, perhaps the finest being *Plant Hunting – Smoke That Thunders* (1927). He became a regular contributor to the *Gardeners' Chronicle* in Britain and to the *Journal of the Arnold Arboretum* and *Horticulture* in America. His advice, however, could be somewhat unorthodox: for example, he recommended using dynamite for preparing tree-holes, rather than simply digging.

When Charles Sargent died in 1927 Wilson became his successor. He had plans to retire to his beloved Gloucestershire, but on 15 October 1930 he and Helen were killed in a road accident when returning from visiting the recently married Muriel. The world was prematurely deprived of an extraordinary botanist with vision, passion, humour and humanity. Wilson is credited with introducing over 1,000 species and there can hardly be a garden that does not contain at least one of these. The impact of his introductions on the early twentieth-century garden is examined in the next chapter, but perhaps it is most fitting that he himself should have the last word:

> Some friends have said 'You must have endured much hardship wandering in out of the way corners of the earth.' I have. But such count for nothing, since I have lived in Nature's boundless halls and drank deeply of her pleasures. To wander through tropical or temperate forest with tree-trunks more stately than gothic columns, beneath a canopy of foliage more lovely in its varied forms than the roof of any building fashioned by man, the welcome cool, the music of the babbling brook, the smell of mother earth and the mixed odors of a myriad flowers – where does hardship figure when the reward is such?

Ernest Wilson's Plant Introductions

The date beside each plant name is the date of its introduction into Britain.

Clematis armandii (1900)

Clematis = From the Greek name *Klematis*,
  for various climbers
armandii = after Père Armand David

Abundant creamy-white, scented 2–3in (5–8cm) flowers in clusters from April to May. A superb, vigorous, evergreen climber with handsome, leathery dark green leaves, strongly ribbed and divided into three leaflets. The variety 'Apple Blossom' has flowers suffused pale pink, with bronzed young foliage, and 'Snowdrift' has profuse pure white flowers.

Ranges from Yunnan to west Sichuan, west Hubei, Guizhuo and Gansu in China, growing among scrub and riverbanks at 200–7,800ft (60–2,400m), spreading up to 20ft 6m.

Acer griseum (1901)

Acer = (Lat.) maple (also means sharp, a reference
  to the hardness of the wood)
griseum = (Lat.) grey

Gorgeous chestnut-red, exfoliating bark is set amid vivid scarlet and orange leaf colours in autumn. The paperbark maple is a lovely, slow-growing, small to medium-sized tree with trifoliate leaves and bunches of 3–5 yellow flowers in May, followed by twin-winged fruits.

Occurs in central China, growing in mixed woods of cherry and maple, reaching up to 50ft (14m) tall.

Davidia involucrata (1901)

Davidia = after Père Armand David
involucrata = having an *involucre*, (Lat.) a ring of
  bracts around several flowers

Delicate, papery white bracts of unequal pairs hang languidly over globular clusters of tiny flowers on long pendulous stalks in May. The romantically named handkerchief or dove tree is a very beautiful medium-sized tree with spreading branches and coarsely toothed green leaves, hairy beneath, which colour orange in autumn. The most commonly grown form is *D. i.* var. *vilmoriniana* (Farges, 1897) with smaller hairless leaves and purplish bark.

From central and western China, growing in mixed forests of maples and beech, reaching 50ft (15m) or more.

Dipelta floribunda (1902)

Dipelta = (Lat.) *di*, two; *pelta*, shield (referring to
  the two shield-like flower bracts)
floribunda = (Lat.) free-flowering

With bounteous fragrant, bell-shaped pale pink flowers, 1¼in (3cm) long and with throats flushed yellow in May, this is a fabulous upright deciduous shrub with deep green leaves and showy winged fruit. Wilson also introduced the related *Kolkwitzia amabilis* (1901) or beauty bush from the same region. This lovely, graceful shrub produces masses of soft pink flowers in May and reaches 8ft (2.5m). He also brought back *Dipelta ventricosa* (1904, lilac-rose flowers), from western China, and George Forrest added *D. yunnanensis* (1910, cream, flushed pink flowers), from north-west Yunnan.

Occurs in west Hubei and Shaanxi in central China, growing in woodland and scrub at 4,000–6,000ft (1,200–1,800m), reaching 17ft (5m) tall.

Viburnum davidii (1904)

Viburnum = Old Latin name for *V. lantana*,
  the wayfaring tree
davidii = after Père Armand David

Handsome, leathery, glossy green, deeply veined leaves are topped with flat heads of white flowers in June. This excellent low-growing evergreen bears ovoid, turquoise-blue fruits in October. Plants are variable and may be either dominantly male or female. Wilson also introduced *V. henryi* (1901, fragrant white flowers and evergreen, fleshy leaves), from west Hubei and west Sichuan in China (first found by Augustine Henry in 1887), and *V. betulifolium* (1901, masses of red currant-like fruits in autumn).

From west Sichuan in China, growing in woodland at 6,000–8,500ft (1,800–2,600m), reaching 3ft (1m) or more.

Lilium regale (1905)
Lilium = (Lat.) lily
regale = (Lat.) royal

Resplendent, trumpet-shaped, powerfully fragrant white flowers, purplish red outside, in July and August. The regal lily is a stunning bulbous perennial, which carries its 1–20 blooms on 3–6ft (1–1.8m) tall stems, clothed in narrow purple-green leaves. From other valleys in western Sichuan in China Wilson introduced the equally beautiful *L. sargentiae*, with very similar fragrant white flowers, greenish-purple outside and broader leaves.

From western Sichuan in China, growing in open, rocky slopes in remote valleys at 2,600–6,500ft (800–2,000m).

Primula pulverulenta (1905)
Primula = from (Lat.) *primus*, first,
  i.e. early flowering
pulverulenta = (Lat.) powdered

Delightful whorls of rich crimson-purple flowers, each with a darker eye, surround powder-covered 3ft (1m) stems in June and July. This beautiful and biggest candelabra primula produces as many as 10 flower-whorls per stem and has crisp green leaves up to 12in (30 cm) long emerging from large, multi-crowned clumps. 'Bartley Strain' is a good paler form, with pink, dark-eyed flowers.

From western Sichuan in China, growing by streams and marshy places above 6,500ft (2000m).

Cornus kousa var. *chinensis* (1907)
Cornus = dogwood; Latin name for cornelian
  cherry, *Cornus mas*
kousa = Japanese name for the species
chinensis = (Lat.) from China

Conspicuous broad white bracts which become tinged pink with age, are borne in fours, each tapering to a neat point and up to 5in (12cm) wide, surround small flower clusters in June. This magnificent large shrub or small tree, with layered branches and deciduous, dark green leaves, bears large, edible strawberry-like fruit in autumn. It differs from the species in lacking tufts of hair under the leaves. The form 'Gold Star' (Japan, 1977) has leaves with a central yellow splash and 'Satomi' has deep pink bracts.

From central and western China, growing in woodland margins and scrub at 4,000–11,000ft (1,200–3,400m), reaching 20ft (6m) tall. The type species occurs in Japan and Korea.

Magnolia sinensis (1908)
Magnolia = after Pierre Magnol, French
  botanical professor
sinensis = (Lat.) from China

Lovely saucer-shaped, lemon-scented, white flowers, 4–5in (10–13cm) wide, with bright crimson stamens, hang from wide spreading branches in May and June. This beautiful deciduous large shrub has deep green leaves, which are densely silky-hairy beneath. Wilson also introduced the very similar *M. wilsonii* (1908, smaller, scented flowers and narrower leaves), and which comes from moist forests in north Yunnan and west Sichuan in China, reaching 26–30ft (8–10m).

From north-west Sichuan in China, growing in forest and thickets at 6,500–8,500ft (2,000–2,600m), reaching 17ft (5m) or more tall.

8

A FORREST OF RHODODENDRONS

George Forrest

(1873–1932)

FROM HIS EARLIEST CHILDHOOD GEORGE FORREST LOVED THE outdoors. As a small boy, like David Douglas before him, he would spend days roaming the Scottish countryside near his home in Kilmarnock, revelling in its wildness and learning the secrets of its natural history. This compulsion to live and work in the open lasted throughout his life, yet it was this spirit of freedom that denied future generations his wisdom and knowledge. For despite repeated pleas from friends and colleagues, Forrest did not write a full account of his twenty-eight years of plant hunting in western China. He repeatedly said that this was to be a job for his retirement, but unfortunately he was never to experience a quiet retirement, and, given his restless nature, who can tell

Left: On his sixth trip in 1924, sponsored by the Rhododendron Society, Forrest brought back the profusely flowering *Camellia saluensis*.

Above: George Forrest: a resourceful Scot, who made the beautiful mountains of Yunnan his own.

whether he could in fact have sat down for long periods to write his memoirs? Forrest died on 5 January 1932 aged only fifty-nine, possibly worn out by the exertions of his seven expeditions. However, he died in a way he would have chosen – having successfully completed his last trip, he collapsed among the wonderful scenery near Tengyueh (now Tengchung) while out shooting, one of his favourite pastimes.

George Forrest was born on 13 March 1873 in Falkirk, and after an education at the Kilmarnock Academy he was employed by a pharmaceutical chemist. He was taught the medicinal properties and uses of many plants as well as simple surgical procedures, and as part of his training he also learned how to dry, label and mount herbarium specimens, a skill that was to prove useful in later life. It appears that he was relatively happy in his job, but when he came into a small inheritance he packed his bags and, using the excuse of visiting relatives, set off to Australia in 1891. He arrived at the peak of gold fever, and decided to try his luck as a prospector; by all accounts he flourished in the harsh conditions and even made a small profit. At the age of eighteen he was displaying his ability to survive, his fortitude, determination and an indefatigable spirit, which later made him such a successful plant hunter.

After a period spent on a sheep farm, Forrest returned to Britain in 1902 after a stop-over in South Africa. There are two versions of how he came to be employed by the Royal Botanic Garden, Edinburgh. The first, more mundane, account states that he simply wrote to Professor Sir Isaac Bayley Balfour, the Regius Keeper, asking if there were any vacancies. The more romantic story has it that one day Forrest was out fishing on Gladhouse Lock when he was forced to seek shelter from a torrential downpour. Looking around him he noticed the corner of a stone coffin poking out of a tumulus. On further inspection the coffin was found to contain a skeleton. Fascinated by his find and wishing to know more, Forrest visited the Antiquarian Museum in Queens Street, Edinburgh. Here he became friends with the museum staff, which led to his meeting the Professor.

The only job vacancy at the Botanic Garden was in the herbarium, and Forrest took this 'until something better came into view'. For the next two years he improved his knowledge of flowering plants and continued to exhibit some of his more eccentric character traits. The office-based nature of the job did not suit his temperament, so to counter this he refused to use a seat while at work and lived outside the city, making the daily 12 mile round trip on foot. Weekends would see him up in the hills, walking, fishing or shooting. In his late twenties the stalwart Forrest made an interesting figure, dressed nearly always in knickerbockers and 'never was he seen in plus-fours'. The impression he gave acquaintances was that he did not suffer fools gladly: he knew what he wanted and where he was going, and had the self-confidence to get there. He had a rich sense of humour and was a fount of anecdote, despite being rather reserved and comfortable only in the company of friends.

Forrest's big break came in 1903, when Professor Balfour was asked by Arthur Kilpin Bulley (1861–1942) to recommend to him a plant hunter to travel to China. Bulley, a wealthy merchant, was excited by the new plants pouring in from China and wanted a collection for his new garden at Mickwell Bow, Ness, near Neston (the gardens are now the Liverpool University Botanic Gardens). Bulley had already been in contact with Augustine Henry but had not been successful in acquiring specimens, so he decided to take the matter into his own hands and became the twentieth-century's first and last sole patron of grand-scale plant hunting expeditions. The arrangements between Bulley and Forrest continued to operate until May 1904, when Forrest left with instructions to explore south-east Tibet and north-west Yunnan in western China. In addition, although Forrest worked for various employers, he maintained a bond with Professor Balfour, sending him extensive herbarium collections from his various expeditions.

Three months after his departure, Forrest arrived at Tengyueh in Yunnan. The British Consul here was George Litton, who proved to be a very helpful and lively companion. The landscape in this region of China is dominated by the three great rivers: the Mekong, the Yangtze and the Salween. All rise in the highlands of central Tibet and cut parallel courses through the landscape before dispersing to flow into three different seas, thousands of miles apart. The reason this area is particularly rich for plant hunters is because the deep valleys and high spurs cut by the rivers and their many tributaries are ecologically isolated and each has developed its own distinct flora. In addition, the region's geology and topography mean that the range of plant types is very wide: the valley hillsides yield many shrubs, the plateau grassland abounds with herbaceous species, and on the high limestone ridges of the Shweli-Salween Divide rhododendrons flourish. To the gardener this last fact may seem somewhat peculiar, but unlike alkaline soil the 'lime' in limestone is locked up in the rock and not available to the plant, so cannot cause problems to ericaceous plants. Forrest was the first to record this phenomenon, and it took many years for his observations to become accepted.

Left: This liana cane bridge across the Yangtze river in Yunnan province was typical of the precarious way in which rivers had to be crossed. The photograph was taken in 1905.

Right: George Forrest: a photograph taken in the field, showing him in typical dress of tweed jacket and knickerbockers.

Forrest had great reserves of self-discipline and good organizational powers, and both proved vital to the success of his expeditions. Although his prime role was that of a plant hunter, in many ways he was a typical example of the Victorian naturalist. Like Sir Joseph Hooker before him, he gathered specimens of mammals, insects, birds (including thirty new species) and geological samples. He set out from Tengyueh, arriving in Talifu (now Dali) in late August 1904. This, the largest town in the province of Yunnan, was Forrest's main base for his exploration of the mountains to the north and west. Due to the lateness of the season, however, he spent much of his first year becoming acclimatized to the country and people and improving his knowledge of the region's flora. He respected and befriended the local people, learning Chinese to a relatively high standard and using his pharmacological knowledge to help cure their ailments. At his own expense, he arranged for thousands of Yunnan Chinese to be vaccinated against smallpox. The Chinese collectors he trained to help him became loyal friends, who respected his honesty and hard work.

In the summer of 1905 Forrest undertook his first serious plant hunting expedition to the north-west corner of Yunnan. The wild landscape, where few Europeans had been before, was both arduous and dangerous: the passes between valleys were blocked with snow for half the year, the rough paths often teetered on the edge of precipices, and the only way across the unnavigable rivers was by rickety bamboo rope bridges. To make matters worse, the whole area was in a state of political turmoil and civil unrest. The lamas (Tibetan priests), who used force and deceit to rule the numerous poverty-stricken and superstitious local tribes, were engendering a strong feeling of xenophobia among them. Enraged by the unauthorized invasion of the Tibetan Holy City of Lassa (now Lhasa) by the British under Colonel Younghusband, and by the Chinese authority's attempts to control the strategic town of Patang (now Batang), the lamas were in the process of extracting revenge. First the Patang lamas rose in revolt and murdered the high-ranking Chinese Government Official and all his followers. They then executed all the French missionaries stationed in the town, along with all their converts, and destroyed the mission's buildings. The insurgence spread to Atuntze, a trading station on the Chinese–Tibetan border. Chinese troops were sent in to quell the uprising but soon found themselves beleaguered by angry locals.

Overleaf: The terrain of Dali (formerly Talifu) in China, with its broad rivers and high hills, made travelling difficult, but the botanical rewards were worth all the effort.

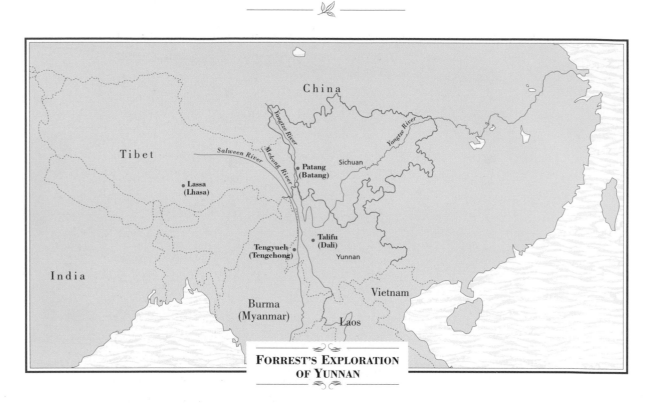

FORREST'S EXPLORATION
OF YUNNAN

Meanwhile, oblivious to all this, Forrest was quietly plant hunting in the hills just three days' walk to the south of Atuntze. He was staying as the guest of Father Dubernard, the chief of the French Catholic mission station at the small village of Tzekou. Situated on the right bank of the Mekong at about 5,000ft, the mission was home to two elderly priests and a few local families who had converted to Christianity. Rumours of atrocities perpetrated by the lamas soon began to reach the village, spreading confusion and terror among Tzekou's inhabitants. Although the mission was well established, everyone was acutely aware of the antipathy the lamas felt towards it. On 19 July the dreaded news arrived – the Chinese troops at Atuntze had been wiped out and a war party of lamas was heading towards the mission. With no means of defending themselves, there was no option but to leave immediately for the nearest friendly village, Yetche, some 30 miles to the south. Following the two priests and Forrest on the moonlit path were his assistants and about sixty men, women and children from the mission. They were in trouble almost as soon as they had set out. The track took them close to the lamasery at Patong and, as they tried to creep by silently, a member of the group made a noise. A shrill whistle immediately echoed around the valley, ending all hope of a covert escape. The

184

party kept up a desperate pace, but early the next morning a villager informed them that the lamas had run through the night and were now ahead of them, creating a blockade.

At midday the group reached a patch of high ground, from where they could see a great cloud of black smoke rising above the mission. The two priests watched in horror as their home was burned to the ground, and the last vestiges of spirit left them as they climbed down the hill. Forrest was anxious to press on as fast as possible in the hope that they could break through the lamas' trap before it was fully set, but the priests had fallen into a state of despair and called a halt beside a stream in the valley. Forrest watched in exasperation as they sat down with their converts to eat some of their provisions and to prepare themselves for death. Feeling the need to take some sort of positive action, Forrest left the group and climbed a nearby spur to reconnoitre the surroundings. No sooner had he reached the peak than he spotted a large body of armed men running down the path they had just descended. He shouted a warning and the panic-stricken party scattered in all directions in a vain attempt to escape. It was too late: the lamas were closing in on their victims. One of the priests, Father Bourdonnec, ran towards the forest in the south of the valley, ignoring Forrest's yells to change direction. He had not covered 200 yards:

> ere he was riddled with poisoned arrows and fell, the Tibetans immediately rushing up and finishing him off with their huge double-handed swords. Our little band, numbering about 80, were picked off one by one, or captured, only 14 escaping. Ten women, wives and daughters of some of our followers, committed suicide by throwing themselves into the stream, to escape the slavery and worse, which they knew awaited them if captured. Of my own 17 collectors and servants only one escaped.

Forrest found himself in a terrible predicament. The valley measured 4 miles long and 1.5 miles wide. To the west it was blocked by a high mountain range and to the east by the fast-flowing Mekong river. The heavily wooded ridges to the north and south were now teeming with the bloodthirsty lamas and their followers. Forrest decided to make a break to the east and dashed down a precarious cliff path of branches and slippery logs:

> On I went down towards the main river, only to find myself, at one of the sharpest turns, suddenly confronted by a band of hostile and well-armed Tibetans, who had been stationed there to block the passage. They were distant about a hundred yards, and sighting me at once gave chase. For a fraction of time I hesitated; being armed with a Winchester repeating rifle, 12 shots, a heavy revolver and two belts of

cartridges, I could easily have made a stand, but I feared being unable to clear a passage before those whom I knew to be behind me arrived on the scene. Therefore I turned back, and after a desperate run, succeeded in covering my tracks by leaping off the path whenever I rounded the corner. I fell into dense jungle, through which I rolled down a steep slope for a distance of two hundred feet before stopping, tearing my clothes to ribbons, and bruising myself most horribly in the process. I then got behind a convenient boulder and made every preparation for a stand should they succeed in discovering my ruse, which I never doubted but they would.

Luck was with him and the Tibetans passed him by. Forrest remained in his hiding place until nightfall, then tried to escape out of the south end of the valley. He climbed up over rock and through dense forest to a height of 3,000ft, only to find his path blocked by groups of lamas with hunting dogs standing watch by campfires. As dawn began to break, he returned dejected to his hiding place. This was to be the pattern for the next eight days and nights. He spent the daylight hours resting up in secluded spots and evading his persistent hunters, and the nights trying to break through the southern barrier. For the whole of this time he was forced to subsist on just one handful of wheat and dried peas, which he had found lying on the ground and had the presence of mind to pick up. On the second day he had to bury his boots in a stream bed because they were leaving a trail of distinctive imprints:

Another day I had to wade waist deep for a full mile upstream to evade a party who were close on my heels; once a few of them came on me suddenly and I was shot at, two of the poisoned arrows passing through my hat; another time my hiding place was discovered by a Tibetan woman, one of many who had been sent out to track me down. Once as I lay asleep under a log in the bed of the stream, exhausted by my night's fruitless journey up the mountain side, I was awakened by the sound of voices, and a party of 30 Lamas in full war paint crossed the stream a few yards above me. Armed as I was I could have shot down most of them, but, though enraged as I was at the time, I held myself in check as I knew that to fire but one shot would be to bring a hornet's nest about me. My only chance was to keep still.

Exhaustion and lack of food eventually took their toll, and Forrest began hallucinating at the start of the ninth day. When he had recovered from his delirium, he realized that physically he could not carry on much longer and that the time had come for a spirited last stand. In the centre of the valley stood a small cluster of huts belonging to the local Lissoo (now Lissu) tribe. That evening the bedraggled plant hunter, covered in cuts and bruises and walking gingerly on swollen feet, approached the village with the intention of obtaining food by force. Fortunately the villagers were friendly and he was not required to use his firearms. The only food they had was a coarse barley dish, but Forrest was so famished that he gorged himself on it. After so long without any sustenance his stomach quickly became inflamed and remained painful for several months. At great risk to himself the chief of the village agreed to help him, and after four days' recuperation took him down to the outskirts of another friendly village. As a Tibetan party hunting for Forrest had spent the previous night in the village, the Scot was hidden in a farmhouse a mile or so away while arrangements were made for his escape. The journey out of the valley was almost as traumatic as his days on the run, and he recorded that 'the misery of all is entirely beyond my powers of description'. The plan was simple – to climb high up into the western mountain range and then head south, skirting the danger zone. However, it was the middle of the rainy season and Forrest and his guides became completely drenched as they trudged up the lower flanks of the mountains. Almost as distressing for the avid plant collector was the fact that they were 'tramping over alps literally clothed with Primulas, Gentians, Saxifrages, Lilies, etc., for these unknown hillsides are a veritable botanist's paradise'. After cutting their way through stands of bamboo and rhododendron forests, they finally reached the snowline at about 18,000ft. For the exhausted Forrest conditions were not about to improve:

> We had no covering at night; no food but a few mouthsful of parched barley, and the rain and sleet fell in such deluges that to light a fire was impossible. On reaching the summit we turned south, travelling in that direction for six days, over glaciers, snow and ice, and tip-tilted, jagged, limestone strata, which tore my feet to ribbons.

By taking this tortuous route Forrest successfully avoided the lamas, but when the party descended over more sharp rocks to the cultivated plateau at 9,000ft, 'to put the finishing touch to my misery', he stepped on a vicious bamboo spike. (The villagers hid these fire-hardened and sharpened pieces of bamboo around their fields to protect their crops from marauders.) The spike, measuring 1 inch in breadth, passed straight through the unfortunate man's foot and stuck out over 2 inches on the topside. Forrest was left in excruciating pain and with a wound that would take months to heal.

Hobbling on in a terrible state, Forrest managed to reach Yetche without further incident. After tending his wounds, eating a hearty meal and enjoying a rest, he set off for the Chinese town of Hsiao Wei Hsi. Following Fortune's example, he disguised himself as a Tibetan to avoid unwanted attention. At

Hsias Wei Hsi he joined a company of 200 Chinese soldiers and made the nineteen-day journey to Talifu, arriving at the end of August. During his nightmarish period of evasion and escape Forrest had been reported as missing, presumed dead. The Foreign Office had sensibly withheld this information for as long as possible in the hope that he had somehow survived: 'thus my family mourned my loss only for a week.'

At Talifu, Forrest was saddened to learn of the violent end met by Father Dubernard. Two days after the massacre in the valley he had been discovered hiding in a cave by the lamas. They dragged him out, broke both his arms and tied them behind his back, and hauled him back to the burned-out mission at Tzekou. They strapped him to a post and for three days systematically tortured him to death. His nose and ears were cut off, his eyes gouged out and his tongue extracted. Each day a joint of his fingers and toes was cut off and finally, at the point of death, he was disembowelled, then beheaded and cut into quarters. Parts of his body and of Father Bourdonnec's were distributed to the various lamaseries in the area, where they were displayed as trophies.

Forrest now had time to reflect on his escape. While grateful to be alive, he lamented the death of so many friends. He was also upset by the loss of all his possessions and of:

> nearly all the results of a whole season's work, a collection of most valuable plants numbering full 2,000 species, seeds of 80 species, and 100 photographic negatives. It is difficult to estimate the value of such a loss; coming from an entirely unexplored area, probably one of the richest in the world, there was undoubtedly a very large percentage of new species. I had sent scraps of specimens home in my letters, and about a dozen of those, or one-third of the number, proved to be new species.

Such an experience would have discouraged a more faint-hearted man, but the resilient Forrest was made of stern stuff. After just a few days of rest he set out to try to recoup his losses. He accompanied his friend George Litton to Tengyueh, and on 11 October 1905 they set off with a large group of porters for the upper Salween district. It was a testing trip for Forrest to make so soon after his previous tribulations. The narrow paths, often clinging to the side of an abyss and choked with vegetation (including a type of stinging nettle the size of a laurel), were dangerously slippery and uneven: 'In some places we had to haul ourselves over boulders by pendant branches, or scramble along the face of the cliffs by notches in the rocks, more suitable for monkeys, Lissoos, or other creatures gifted with more prehensile feet than a European.'

There was scant evidence of animal or bird life to restock their food supplies, and because the sheltered valleys had an almost tropical microclimate he discovered that the:

> insect life is both vigorous and troublesome. Creatures with inconveniently long legs plunge suddenly into one's soup; great caterpillars in splendid but poisonous uniforms of long and gaily coloured hairs arrive in one's blankets with the business-like air of a guest who means to stay. Ladybirds and other specimens of *Coleoptera* drop off the jungle down one's neck, whilst other undesirables insert themselves under one's nether garments. The light in the tent attracts a perfect army of creatures which creep, buzz, fly, crawl and sting. Scissor insects make the day hideous with their strident call, and the proximity of Lissoo coolies introduces other strangers, of which *Pulex irritans* [the human flea] is by far the least noxious.

The beauty of the country more than compensated for any discomforts. Later Forrest would wax lyrical about the Salween scenery, exclaiming that it:

> can never be forgotten by anyone who has wondered at it in the rich sunshine which prevails after the autumn rains have given way to the first touch of winter. The great variety of rock formation, the abundant forests and vegetation, and the diversity of light effects between the summits of the ranges (at 10,000 to 13,000') and the abyss in which the river flows produce a vast panorama of ever-changing beauty. In the morning, the sun, as it touches the top of the Mikong divide, sends wide shafts of turquoise light down the side gullies to the river, which seems to be transformed into silver. The pines long the top of the ridges stand out as if limned by the hand of a Japanese artist. In the evening all the wide slopes of the Mekong side are flooded with red and orange lights, which defy photography and would be the despair of a Turner.

Although the party had been warned about the bellicose nature of the local inhabitants (a group of Germans had been massacred a few months earlier) and saw much evidence of intertribal warfare, they managed to avoid any serious confrontations. One surreal encounter was with a group of warriors led by a prophet clutching a piece of paper. The prophet informed them that he had received instructions from Heaven to kill someone, and thought that the chief of the neighbouring tribe would make a suitable victim, but he would appreciate the foreigners' opinion on his choice. Somewhat taken aback, Forrest 'strongly recommended him to go home and see to the grinding of his maize'. On another occasion they inadvertently became embroiled in a dispute concerning the ownership of a rope bridge. The villages on either side of the river both claimed responsibility (and therefore both wanted the toll money),

Forrest's camp in the Lichiang Mountains in north-west Yunnan. In the foreground
are several of the team who helped Forrest throughout his explorations.

but when the chief of the opposite bank let loose a poisoned arrow, Forrest and
Litton decided to take sides. They fired their Winchester rifles at the cliff next
to the chief, who, on seeing the rock exploding around him, decided discretion
was the better part of valour. The expedition ended on a sad note when, upon
their return to Tengyueh, Litton was struck down with malaria and died of
blackwater fever on 9 January 1906.

A melancholy Forrest set off in March to explore the Lichiang (now Likiang)
mountain range where it forces the Yangtze river into a mighty curve. He spent
several fruitful months scouring this 'one vast flower garden' with his team of
trained plant gatherers, but was struck down by the Salween malaria which had

190

been lying dormant in his system. He had to travel back to Talifu to seek medical help, but once there he ignored the doctor's pleas for him to return to Britain. Fear of losing another collecting season overcame his concern for his own health, and he spent numerous painful weeks recuperating while simultaneously organizing his small army of native workers to collect for him. Eventually Forrest's persistence paid off, and he sailed home in 1906 with an impressive hoard of seeds, roots and plants, including *Lilium langkongense*, *Rhododendron dichroanthum*, *Primula bulleyana* and *P. vialii*.

Despite the traumas of his earlier trips, Forrest made a further six excursions to his beloved Yunnan over the next twenty-six years. None of the subsequent visits were quite as perilous as his first, but he was still hampered by local hostility and civil unrest. Living in China was, wrote Forrest with feeling, like camping on an active volcano. On his third trip he arrived in Tengyueh in 1912 to find the country in uproar. A revolution had broken out the previous autumn which soon spread to all districts, causing chaos and confusion. Old scores were settled in the typical violent manner of the area, and new atrocities were being committed in the so-called name of freedom. Forrest was forced to retreat across the border into Burma, from where he sent out his band of locals to collect for him. After a few months of forced inactivity, he was able to return to Yunnan and recommenced plant hunting in the Tali and Lichiang ranges. Among the many other botanical delights he encountered during this time was the magnificent *Rhododendron sinogrande*. He returned to Talifu in October 1913, but on 'the day I entered the city, the local soldiery, some 3,000 troops,

The orchid-like *Primula vialii* is one of the most striking of the perennials that Forrest brought back from his first trip to Yunnan.

Despite the revolution that plagued his third trip to Yunnan in 1912, Forrest found many gems, including the magnificent *Rhododendron sinogrande*.

mutinied, shot down their officers when on morning parade, and captured the city'. Forrest and his friend the Reverend Hanna were taken prisoner and forced to work as medical assistants to the local militia for three weeks. Talifu was eventually stormed by loyal troops from the provincial capital, Yunnanfu, and after much bloodshed the city was recaptured.

Forrest's fourth trip of 1917–19 and his subsequent ones (1921–2, 1924–5 and 1930–32) were sponsored by the Rhododendron Society, which sent him out in response to the growth of rhododendromania. He was so successful that he collected over 300 new species, including the scarlet *R. griersonianum*. This wealth of new material required the revision of the *Rhododendron* genus, a task undertaken by Forrest's mentor, Professor Balfour. During these years other plant hunters, including Frank Kingdon-Ward, were also exploring north-west Yunnan, much to Forrest's annoyance. Although he did not hesitate to follow in their footsteps, he became irritated when they encroached on his 'territory'. As time passed and Forrest slipped into middle age, he would announce at the start of each expedition that it was to be his last. In November 1930 he embarked on his 'final' trip with the intention of visiting all his old haunts. He was particularly keen to collect those specimens that had not flourished when first introduced to Britain. It was a highly successful season, and in one of his last letters back home he triumphantly wrote:

> Of seed, such an abundance, that I scarce know where to commence, nearly everything I wished for and that means a lot. Primulas in profusion, seed of some of them as much as 3–5 lbs., same of Meconopsis, Nomocharis, Lilium, as well as bulbs of the latter. When all are dealt with and packed I expect to have nearly, if not more than two mule loads of good clean seed, representing some 4–500 species, and a mule-load means 130–150 lbs. That is something like 300 lbs. of seed … If all goes well I shall have made a rather glorious and satisfactory finish to all my past years of labour.

It was a fitting end to a remarkable career. Forrest died of a heart attack on 5 January 1932 and was laid to rest in the little churchyard at Tengyueh, close to his old friend George Litton. It is a suitable site for a man whose heart always lay in the wild mountains of the Himalayas.

If the wealth and beauty of the flora of the Far East astounded plant hunters such as Fortune, Forrest, Wilson and Kingdon-Ward, the impact of the plants on their return to Britain was no less profound. However, in terms of changes to the garden, it is necessary to gather their work together. After over half a century in which artifice dominated the garden there came the inevitable backlash – nature, which for so long had been banished, was once more welcomed. The culmination of the rejection of Victorian excess was the emergence of the Edwardian garden.

The fiery Irishman William Robinson (1838–1935) was the most outspoken advocate for the return of nature and hardy plants to the garden. In 1883 *The English Flower Garden* was first published, and Robinson's writings encouraged the change of emphasis in planting design. The Victorian fashion for artistically arranging plants gave way to the grouping of plants to create a scenic display. This notion extended into the conservatory, where tiered shelves were removed and specimens taken out of their pots and planted in beds. An example of this type of display can still be seen in the Palm House at Kew.

In the landscape, conifers fell from favour and deciduous trees returned. Many of the new trees and large shrubs from the Orient were welcomed on account of their interesting form, exotic flowers or colourful foliage. A particular fascination developed for variegated, purple and golden-leaved foliage forms. Below the trees Oriental herbaceous plants, bulbs and small shrubs were added. The 1880s saw a new trend – the creation of replica foreign scenery designed to look as natural as possible. A particularly popular theme was the Himalayan rhododendron forest, inspired by Hooker's *Himalayan Journals*, exemplified at Cragside in Northumbria, where Lord Armstrong had planted 'several hundred thousand' rhododendrons by the 1890s.

As the nineteenth century drew to a close, Robinson's dogmatic proclamations became too much for the architects John Dando Sedding and Sir

Reginald Blomfield, who rose to defend a more formal approach to garden design. This has become known as the Battle of Styles, and as the acrimonious debate raged, a new style emerged – the Arts and Crafts vernacular garden based on a return to Englishness.

The years before the First World War were the last period of great wealth accumulation and great house building, and it was Sir Edwin Lutyens (1869–1944) who more than any other architect came to epitomize the Edwardian expression of Arts and Crafts vernacular architecture. His skill, the symbiosis of site with local materials and with

Forrest was captivated by the azure blue flowers of *Gentiana sino-ornata*, perhaps south-west China's prettiest alpine, which he discovered in 1910.

traditional techniques, was without comparison. The boom in house building was matched by one in garden-making. The acknowledged leader of the new garden movement was Gertrude Jekyll (1843–1932). Miss Jekyll used the principles of colour theory to paint garden pictures with plants. As Brent Elliott observed in *The Country House Garden* (1995): 'The first of Jekyll's innovations to become widely accepted was in the planting of the mixed or herbaceous border … Borders were traditionally planted in rows or blocks: Jekyll recommended planting them in large, irregular masses instead, so that the plants would be grouped in pseudo-natural drifts.'

Miss Jekyll and Lutyens were kindred spirits, and their houses and gardens became, and still are, much sought after. Together they created a new kind of garden. The garden framework was an extension of the house's architecture and the layout depended upon inventive geometry, with division and enclosure provided by wall or clipped yew hedge. The garden was always filled with planting which was simultaneously disciplined and profuse, with hardy flowering plants carefully arranged to take into consideration seasonality, colour, form and scent. Not everyone could afford a Jekyll garden, but through her numerous books and her frequent articles she spread her message. As a result a new garden fashion arose, one by which we are still strongly influenced today. Within this type of garden the new hardy plants from the Orient were given a warm and friendly welcome by the more 'old-fashioned' residents. Rhododendrons, conifers and ornamental trees and shrubs graced the woodland and wild garden, while alpines nestled among the nooks of the rock garden and the crannies of the dry stone wall. In the formal garden around the house, the beds and borders were enlivened with new annuals and biennials, bulbs and herbaceous species, all of which proudly held up their glorious heads and dazzled the visitor.

The Lutyens/Jekyll partnership epitomizes the Edwardian garden – 'Gardens of a Golden Afternoon', as Jane Brown, in her book of that name, aptly described them. But in 1914 the sun disappeared, never again to shine so brightly. The country house and garden, with its high labour requirements and dependence on low fuel costs and taxes, was ill-suited to adapt to the enormous social and economic changes wrought by the First World War.

George Forrest's Plant Introductions

The date beside each plant name is the date of its introduction into Britain.

Primula bulleyana (1906)

Primula = from *primus*, (Lat.) first, i.e. early
   flowering
bulleyana = after Arthur Killin Bulley of
   Bees Nurseries

Colourful whorls of soft orange flowers, red in bud, are borne on 28in (70cm) stems in June and July. A beautiful candelabra primula, with large leaves up to 14in (35cm) long. Forrest also introduced *P. beesiana* (*c.* 1906, rose-carmine, orange-eyed flowers) and the lovely *P. poissonii* (*c.* 1906, purplish-crimson flowers).

Native to north-west Yunnan and southern Sichuan in China, growing in marshy mountain meadows and by streams at 8,200–10,000ft (2,500–3,300m).

Primula vialii (1906)

Primula = from (Lat.) *primus*, first, i.e. early
   flowering
vialii = after Paul Vial, French missionary

Exquisite, orchid-like spires of small, fragrant, lilac-pink flowers contrast with bright scarlet buds on 12–24in (30–60cm) stems in July and August. A quite singular species, with narrow, softly hairy leaves 8–12in (20–30cm) long, which appear in May, later than any other species. Originally discovered by Delavay *c.* 1885 but not introduced. Forrest brought back many lovely primulas, including *P. secundiflora* (reddish-purple, bell-shaped, nodding flowers) and his namesake *P. forrestii* (bright yellow flowers and of difficult cultivation).

Occurs in north-west Yunnan and south-west Sichuan, China, growing in damp meadows at 2,800–3,350m.

Gentiana sino-ornata (1910)

Gentiana = after King Gentius of Illyria, 500 BC
sino-ornata = (Lat.) *sino*, China, *ornata*,
   ornamental or showy

Sensational 2in (5cm) wide, azure blue trumpets erupt from long pointed buds in floriferous drifts over mats of trailing, bright green stems and leaves in autumn. This is a fabulous alpine gem. Several forms exist, including 'Praecox' (early flowering) and 'Alba' (white).

Occurs in south-west China on rocky hillsides and meadows at 12,000–18,000ft (3,600–5,500m).

Rhododendron haematodes (1911)

Rhododendron = (Gk) *rhodo*, rose; *dendron*, tree
haematodes = (Gk) blood-like colour

Intense scarlet to crimson bell-shaped flowers in May and June. A splendid, compact, slow-growing evergreen with dark green leaves, covered beneath by thick, woolly red-brown felt. Good hybrids include 'May Day' (× *R. griersonianum*, 1932), bright red, and 'Humming Bird' (× *R. williamsianum*, 1933), cherry-red. Forrest also introduced *R. h.* ssp. *chaetomallum* (1918, blood-red flowers in March and April) and the beautiful *R. neriiflorum* (1910, scarlet-crimson flowers).

Native to western Yunnan in china, growing in alpine meadows and scrub at 11,000–13,000ft (3,400–4,000m), reaching 2–4½ft (60cm–1.5m) tall.

Rhododendron sinogrande (1913)

Rhododendron = (Gk) *rhodo*, rose; *dendron*, tree
sinogrande = (Lat.) *sino*, China; *grande*, showy, big

Enormous, shining, oblong dark green leaves, felted silvery-grey beneath, bear appropriately huge trusses of creamy-yellow, bell-shaped flowers with a crimson blotch in April. This magnificent small tree is the foliar champion of the genus and can have leaves of up to 3ft (90cm) long and 12in (30cm) wide. Forrest also introduced *R. protistum* (1919), a gigantic species, soaring to 98ft (30m) tall, with huge 22in (55cm) leaves and rose-pink to crimson-purple flowers in February to March.

Occurs in Yunnan in China, to Upper Burma and west to Arunachal Pradesh and south-east Tibet, growing in mixed forest and rainforest, with rhododendrons and bamboo, at 6,900–14,000ft (2,100–4,300m), reaching 40ft (12m) tall.

Rhododendron griersonianum (1917)

Rhododendron = (Gk) *rhodo*, rose; *dendron*, tree
griersonianum = after R.C. Grierson, Yunnan
  Customs

Blazing geranium-scarlet bell-shaped flowers top branches of slender dark green pointed leaves in June. This striking free-flowering species of open habit and sticky young growth is a key species in rhododendron breeding, with a vast progeny. Its innumerable hybrids include: 'Elizabeth' (× *R. forrestii* Repens, 1939, deep red), 'Fabia' Group (× *R. dichroanthum c.* 1930, scarlet), 'Tortoiseshell' (× *R.* 'Goldsworth Orange', 1945) and 'Vulcan' (× *R.* 'Mars', 1938).

Occurs in western Yunnan in China, and adjoining Upper Burma, growing in open glades and scrub in conifer and mixed forest at 6,900–9,000ft (2,100–2,700m), reaching 10ft (3m) tall.

Camellia reticulata (1924)

Camellia = after Georg Josef Kamel,
  a pharmacist from Brno
reticulata = (Lat.) netted, net-like pattern

Gorgeous 4in (10cm) wide rose-pink single flowers, with bright yellow stamens in April and May. This superb evergreen shrub or small tree has handsome leathery leaves, attractively net-veined. The semi-double form 'Captain Rawes' (1820) was long thought to be the type species before this wild form was discovered. Tender cultivars include 'Trewithen Pink' and 'Mary Williams' (Caerhays, 1942), and it has also been used in breeding, e.g. 'Salutation' (× *C. saluensis*, soft pink), and 'Leonard Messel' (× *C.* × *williamsii*, rich pink).

Occurs in western Yunnan in China, growing in pine forest and scrub at 6,500ft (2,000m), reaching 30ft (10m) (only 13ft (3m)+ in gardens). It has long been cultivated in China, especially around temples.

Camellia saluensis (1924)

Camellia = after George Josef Kamel,
  a pharmacist from Brno
saluensis = (Lat.) of the Salween river
  of Burma and China

Delicate 2½in (6cm) wide soft pink single flowers smother the rigid dark green foliage in May. This lovely evergreen medium-sized shrub is very important in camellia breeding. It was crossed with *C. japonica* by J.C. Williams at Caerhays Castle, Cornwall (*c.* 1925), and produced the acclaimed *C.* × *williamsii*. These are wonderful free-flowering cultivars, e.g. 'Anticipation' (deep rose), 'Debbie' (pink), 'Donation' (orchid pink) and 'J.C. Williams' (phlox pink).

From north-west Yunnan in China, in the volcanic range of the Shweli river basin and the Yangtze valley, growing in scrub on rocky slopes and by streams at 6,000–9,000ft (1,800–2,700m), reaching 6½–20ft (2–6m) tall.

Magnolia campbellii ssp. mollicomata (1924)

Magnolia = after Pierre Magnol, French
  botanical professor
campbellii = after Dr Archibald Campbell
mollicomata = (Lat.) soft-haired

Sumptuous, pink to rosy-purple, water-lily-like flowers appear on the ends of bare branches from February to March. The hardier Chinese form of the Himalayan pink tulip tree is a medium-sized tree with large broad leaves, often hairy beneath (thus differing from the species), and silver-grey bark.

Occurs in Yunnan in China and north-east Burma, growing in temperate rainforest, with oaks, magnolias (e.g. *M. excelsa*) and rhododendrons (*R. grande*, *R. arboreum* and epiphytic *R. dalhousiae*) at around 7,800ft (2,400m)+, reaching 65ft (20m) tall. The species ranges from east Nepal, Sikkim and Bhutan to south-west China.

9

KINGDOM OF THE BLUE POPPY

Frank Kingdon-Ward

(1885–1958)

O N TWENTY-TWO JOURNEYS SPANNING FORTY-FIVE YEARS, FRANK Kingdon-Ward hunted plants in the remotest corners of Burma, Tibet and Assam. He was also a prolific if unaccomplished writer, publishing some fourteen books in which he clearly took pleasure in sharing his adventures with his readers. He was born on 6 November 1885 at Withington, Lancashire. Earlier that year his father, Professor Harry Marshall Ward (1854–1906), had been appointed Professor of Botany at the Royal Indian Engineering College at Cooper's Hill in Surrey, and ten years later, secured the Chair of Botany at Cambridge. Inspired by his father, Frank acquired a love for nature, except

---

Left: The blue poppy, *Meconopsis betonicifolia.* This was reintroduced by Kingdon-Ward in 1924, and the large quantity of seed he gathered ensured that his introduction was successful.

Above: Frank Kingdon-Ward was the longest serving professional plant hunter, exploring Asia's mountains for forty-five years. He was also a dedicated geographer to whom the Himalayas held an endless fascination.

snakes, of which he was terrified. He was an independently minded child, frequently camping out alone, and from Schrimper's *Plant Geography* he caught the travel bug: 'These illustrations made a deep impression on me, and burnt on my heart a profound desire to see for myself the tropical forest region.'

After attending St Paul's School in Hammersmith, London, Kingdon-Ward entered Christ's College, Cambridge at the age of nineteen to read Natural Sciences, but his father's premature death in 1906 and the poor state of the family finances meant that he was forced to find a job. He rushed his studies, gaining a second class honours degree in two rather than the usual three years, and took the first job that offered overseas travel. He sailed for China in 1907 to become a teacher at the Shanghai Public School. During a two-day stopover in Singapore, Kingdon-Ward finally fulfilled his dream of visiting a tropical forest. He hiked into the interior and walked spellbound through the jungle. When night fell, he slept under the stars.

Further fuelled with the romance of adventure from reading Joseph Hooker's *Himalayan Journals*, Alfred Russel Wallace's *Island Life* and Henry Bates's *Naturalist on the Amazon*, Kingdon-Ward spent his summer holidays exploring Java and Borneo. He was particularly keen to find the pitcher plants described by Wallace, but failed to do so, although he was successful some years later when visiting Singapore and Assam. Back in Shanghai, he approached teaching with a somewhat flippant attitude, regarding it merely as a means to an end. It came as no surprise to his employers when two years later he gave notice in order to travel across China. Malcolm P. Anderson, an American zoologist, had heard of Kingdon-Ward through a mutual friend and invited him to join his expedition. The aim of the trip was to cross central and western China to south Kansu, collecting animal specimens *en route*. The party set off in September 1909, and the inexperienced Kingdon-Ward made himself useful assisting the party. He managed to assemble a small collection of plants, which he later presented to the Botany School at Cambridge.

On his return to Shanghai a year later, he resumed his teaching job, but 'travel had bitten too deeply into my soul' and he became impatient to be off again. As luck would have it, in January 1911 he received a letter from Arthur Kilpin Bulley asking him to undertake a plant hunting expedition to Yunnan. Bulley's previous collector, George Forrest, had been 'poached' by J.C. Williams of Caerhays in Cornwall, and Bulley needed to find a replacement. As a novice plant hunter, with a lack of field experience, Kingdon-Ward was to undergo a baptism by fire.

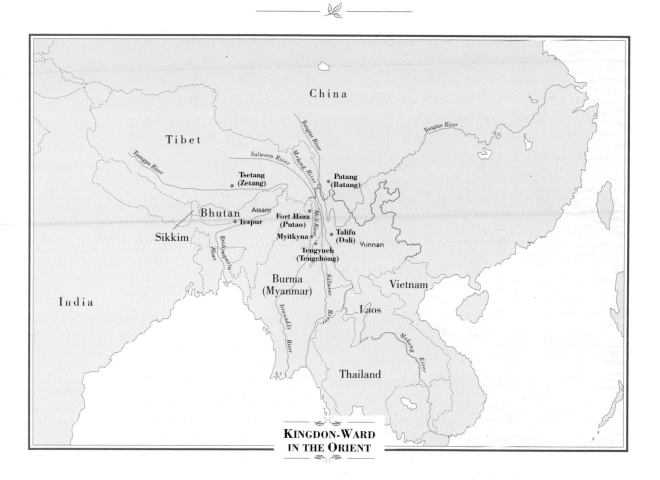

**KINGDON-WARD
IN THE ORIENT**

He left Shanghai in early 1911 for the first stage of his expedition, the lengthy journey to his base at Talifu (now Dali) in Yunnan. After two boat trips, a train journey and a slow steamboat ride, his mule train finally trotted up towards the frontier with China on 26 February. As he left civilization behind, Kingdon-Ward began to realize the enormity of his task and felt acutely anxious and lonely: 'Never again did the sense of paralyzing isolation come so vividly upon me as on that first night, when all the trials that awaited me seemed to take shape and rise in arms to mock my ignorance and feebleness.'

On the journey to Tengyueh Kingdon-Ward became lost after exploring a route away from the main trail and suffered the indignity of being thrown three times by his obstinate mule. At Tengyueh he nursed his bruises while a guest of a Mr Rose, who advised him that Atuntze might make a good destination, thus avoiding any encroachment into the area being explored by the jealous George Forrest (the two men met in July 1913 on Bei-ma-Shan). Kingdon-Ward set out

201

one bright early spring morning, riding through hills coloured by gentians, orchids and roses. After two weeks he entered remote valleys of rhododendrons and camellias, primulas and irises. Foreigners were a rare sight, and he always attracted a crowd of onlookers, some of whom took more than a passing interest. Thefts occurred from his tent and he was incensed at the loss of his most prized possession, his tea thermos.

He reached Talifu on 31 March. Here he hired two Chinese, Kin and Sung, who remained with him for the rest of the year. On 7 April the small team left Talifu, travelling north towards the upper reaches of the Yangtze before turning west across to Atuntze near the Mekong river. On the second day of the westward passage over the Mekong watershed, Kingdon-Ward again became hopelessly lost. After a steep ascent to the Li-ti-ping plateau he sent the caravan on and headed off, accompanied by Kin, into the nearby forests to hunt pheasants. Kin felt unwell and returned to the caravan, leaving Kingdon-Ward to wander up through the wood of fir and bamboo alone. He was constantly sidetracked by the floral marvels around him and after a couple of hours of erratic meandering found himself on what he thought was the main trail. Realizing he was wrong, he foolishly decided to save time by following a stream down to the city of Wei-hsi. He plunged headlong into the forest in the hope of finding such a stream, met a dense mass of bamboo, and by late afternoon had become hopelessly disorientated. That night he was forced to sleep out in the open with only his torn raincoat for shelter and a shotgun with one cartridge for protection against wolves. He was without even a match to light a fire.

Awaking in a deep puddle of rainwater after a restless night, Kingdon-Ward headed down towards a large valley he had seen the previous day and soon found a well-worn trail. He followed this for three hours until he reached a viewpoint, only to find he had been heading in completely the wrong direction! Hungry, wet and miserable, he wisely retraced his steps. He resorted to eating rhododendron flowers, which subsequently gave him a violent stomachache (some species are quite poisonous), before shooting a poor finch at point blank range: 'When I came to pick him up, I found that the No. 6 shot had not only killed him but very nearly plucked him as well. With the exception of the feathers, entrails and beak, I ate him entire.'

Kingdon-Ward eventually came upon a sheep pen he had passed the day before. The main path was still some way off and he attempted yet another short cut, only to lose his bearings. At last he reached the main trail and struggled on to Wei-hsi. Arriving after sunset, he knocked on the door of a house:

'an old woman stood before me, a flaming pine torch held high above her head as she peered into the night. "Why it's his Excellency!" she exclaimed in aston-ishment and ran back to bring her husband. My fame had preceded me!'

Kingdon-Ward reached Atuntze by mid-May and had settled in for the sea-son when, in late July, he received word that the British had marched into Lassa (now Lhasa) and that in retribution the Chinese were intent on killing all Englishmen. Deciding that he was very vulnerable on his own, he made the 180-mile journey north to Patang (now Batang), where other Europeans were stationed, arriving in a mild state of panic, only to find it was a false alarm. He returned disgruntled to Atuntze, having wasted three valuable weeks. October brought news of a 'real' revolution in the south, and on 1 November he left Atuntze for Tengyueh. However, he took a westerly route in order to gather seeds along the Salween river. The Chinese authorities had forbidden him to travel in this area, and considering the trouble elsewhere he was pushing his luck. Inevitably, he encountered members of the Revolutionary Army, but to his great relief they proved unconcerned and courteous, even to the extent of assuring him safe passage through the villages.

Exhausted, Kingdon-Ward, reached the safety of Tengyueh in mid-December 1911. He was now a confirmed plant hunter and was wholly under the spell of the wild Chinese landscape, something he made clear in the final paragraph of *Land of the Blue Poppy*:

Convinced as I am that with its wonderful wealth of alpine flowers, its numerous wild animals, its strange tribes, and its complex structure it is one of the most fasci-nating regions of Asia. I believe I should be content to wander over it for years. To climb its rugged peaks, and tramp its deep snows, to fight its storms of wind and rain, to roam in the warmth of its deep gorges within sight and sound of its roaring rivers, and above all to mingle with its hardy tribesman, is to feel the blood coursing through the veins, every nerve steady, every muscle taut.

Although he sent many new species back to Arthur Bulley, for example *Saxifraga wardii*, *Gentiana wardii*, *Androsace wardii*, and species of *Meco-nopsis*, disappointingly few of them survived in cultivation. However, the following year, 1912, he returned to the same area and found, among others, *Rhododendron wardii* growing on the Do-kar-la Pass in fir forest and meadows:

Overleaf: Terraced fields, steep valleys and rushing rivers are typical of the stunning scenery of Yunnan province.

'In some places the scrub thickets were composed entirely of Rhododendrons – *R. Wardii* with lemon-yellow flowers … has flowered in England from my seed, and is highly spoken of, both for its fine foliage and for its flowers.'

In 1914 Kingdon-Ward travelled to Hpimaw in the frontier region of north-east Burma, near the Nmai-Salween Divide, with the intention of travelling north along the mountain chain and then west into Assam. The expedition was unpleasant. Constant rains meant that armies of 'leeches entered literally every orifice except my mouth'; the bites became infected and took weeks or even months to heal fully. He endured recurrent bouts of malaria, and battling through dense bamboo forests wore down his reserves. To add to his misery, a house collapsed on him and he twice fell from precipices. Outrageous luck meant that a tree broke the first fall and a narrow ledge high above a sheer drop stopped the second, and the tenacious Kingdon-Ward battled on, refusing to give up. When he went down with fever on the first attempt to scale Mt Imaw Bum, he returned to climb both it and its neighbour, Lacksang Bum.

When his deteriorating health reached a critical point, Kingdon-Ward had the common sense to abandon the area and move towards Assam, taking a route across the Wu-law Pass and into the Laking valley. Here food shortages forced him to reduce his party for the journey across the Shingrup Kyet Pass to the Mali river valley and north to the fort at Konglu. Rested, cleaned and fed, he set off upriver for Fort Herz (now Putao), arriving on 27 September 1914. The next day, however, he collapsed with malaria and all thoughts of Assam, now just a few miles to the west, were forgotten. After six weeks of convalescence, he was sufficiently recovered to travel down to Myitkyna on the Irrawaddy river for Christmas.

This traumatic expedition made Kingdon-Ward reconsider his chosen vocation when the offer of a permanent botanical post in England was put to him. However, the trip had at least yielded the beautiful *Primula burmanica* and revealed the huge plant hunting potential of the area. He turned down the offer and continued plant hunting, pausing only to enlist in the Indian Army for the remainder of the First World War, rising to the rank of Captain.

Fortunately, Kingdon-Ward never suffered quite as much on his subsequent journeys, and he even returned to the same area in 1919. For his third and last trip to China in 1921, he returned to Yunnan, Sichuan, south-east Tibet and north Burma. The prize of this trip was the slipper orchid *Paphiopedilum wardii* (which he introduced *c.* 1932). After his return to Britain and marriage to Florinda Norman-Thompson, he turned his attention to Burma and Tibet.

In 1924–5 he embarked on his finest journey – to the mysterious region of the Tsangpo river (now Yarlung Zangbo Jiang) in the eastern Himalayas. This region held a great fascination for Kingdon-Ward, who was a keen geographer, and he was to find some of his most celebrated plants here. On its upper course in Tibet the river is slow and navigable, likewise on its lower reaches in Assam, where it becomes the Brahmaputra river. Between lay 50 miles of unknown landscape in which, it was rumoured, were thundering high waterfalls and a fierce torrent hidden within sheer-sided, bottomless canyons.

Kingdon-Ward's intention was to cross through the Sikkim Himalayas into south-east Tibet, spend the season collecting and in the autumn approach the gorges from the north. Despite the three-quarters of a century since Joseph Hooker's visit, Sikkim remained a difficult place to enter. Permits were required for both Sikkim and Tibet, and these were obtained from the Political Officer, Lieut-Col. F. M. Bailey, who had made a preliminary exploration of the Tsangpo gorges and provided valuable information on the region. On this trip Kingdon-Ward travelled with Lord Cawdor, and on 23 March 1924 they crossed the Nathu La Pass into Tibet, a land of austere beauty, still gripped by winter. The mighty spine of the snow-covered Himalayas stretched across the horizon as far as they could see. It was a bleak, harsh land whipped by fierce, dust-laden icy winds. However, as they crossed the frozen, rocky hills of the high desert, their thoughts were on the warmth of the brief summer still some months away. This they hoped to enjoy in the sheltered and forested valleys to the east, where a profusion of hardy flowers waited to be gathered.

The small party continued east, reaching Tsetang (now Zetang) on the banks of the Tsangpo on 21 April. Now they exchanged frozen wastelands for heavy rain as they moved downstream to Tsela Dzong. The hard travel, poor food, damp accommodation and omnipresent insects had taken their toll, and the party remained at Tsela for a few weeks to recuperate before heading off north-east across the Temo La Pass to Tumbatse on the Rong-chu river. Nestling in a green valley, surrounded by fir forests and rocky slopes rich with rhododendrons, this was to be their base for the next five months. On one occasion Kingdon-Ward had pointed out a clump of bright leaves to Lord Cawdor:

> and he taking out his telescope, looked at them long and carefully. 'Why' said he, at length 'they aren't leaves, they are flowers; it's a Rhododendron I believe'. 'What' I shouted, almost seizing the glass from him in my eagerness … on the naked rock was Scarlet Runner in proud isolation … The virgin snow dabbled with its hot scarlet, spread a bloody sheet over the tortured rock.

The marshy meadows were filled with drifts of primulas, including *Primula alpicola* and *P. florindae*, which Kingdon-Ward named after his wife:

> It is the forest of stems rising above the curved leaves, like frail masts from a rough sea, which is the attraction. A stream blocked by colonies of *P. florindae* in flower is for ever a bit of the Tibetan Plateau ... the flowers themselves are so crowded that while they shake themselves free from the mop you get the feeling the remainder must explode like a rocket, and send a cloud of scented yellow stars drifting to earth.

On 9 August the party undertook a five-week excursion up the river Tonkyuk (now Dyongjug) across high passes into unknown territory around Lake Atsa Tso on the Salween divide. These were virgin lands, and in the pine forests Kingdon-Ward found *Lilium wardii*, 'a beautiful shade of pink, closely and evenly spotted with maroon purple'. During the autumn they methodically collected seed of the plants they had found, digging through thick snow if necessary. One of the many skills that made Kingdon-Ward such a successful plant hunter was his exceptional memory for locations. He could remember with absolute accuracy where a particular plant grew, enabling him to return and collect seed months later. Their haul included a large quantity of seed of the fabulous Himalayan blue poppy (*Meconopsis betonicifolia*), first found by Delavay in 1886 in Yunnan, and the easiest of the blue poppies to grow, which Kingdon-Ward had found to be common in the area: 'under the bushes, along the banks of the stream ... grew that lovely poppy ... The flowers flutter out from amongst the sea-green leaves like blue-and-gold butterflies.'

The seed harvest finished in mid-November, and the party began to explore the Tsangpo gorges. Descending into the upper gorge, they found the forest glowing with autumnal tints. Below was the river, which the group crossed in skin coracles. The party stopped at Gyala, the last village before the challenging steeper sections, where 'We could still make our way down to the river-bed, where the rocks looked as if they were simmering like red-hot lava; this on closer examination turned out to be due to layers of *Cotoneaster conspicuus* covered with thousands of scarlet berries.'

The intention was to travel along the Kongbo Tsangpo to a small monastery in the forest and then continue on into the steepest gorge. Leaving Gyala on 16 November with twenty-three men, one sheep and two dogs, they entered forests full of tree rhododendrons and bamboo. The ground was too uneven to pitch their tents, and the canvases had to be rigged to the trees to form a shelter. Piles of bamboo were cut and formed into mattresses, and the following

Primula florindae (giant cowslip) was one of Kingdon-Ward's most successful
discoveries. He named it after his first wife, Florinda.

morning, after an uncomfortable night, Kingdon-Ward wandered into the forest to search for rhododendrons:

> on my way back to camp, I noticed a tangle of rhododendron, the stems rising no
> more than a foot above the ground ... with small thick leaves, and I was sure I had
> never seen it before ... there were flower buds too, the bud scales finely silver-
> fringed with hairs like spun glass.

Right: Frank Kingdon-Ward: portrait of a professional plant hunter and explorer.

Below: Kingdon-Ward's 1933 expedition team rest amid the splendid scenery of south-east Tibet.

He never saw the rare *Rhododendron leucaspis* again. On the fourth day they reached the Pemakochung monastery and took a much needed day's rest. The monastery was in a magical spot, surrounded by a rhododendron forest that obscured the cliffs. In spite of the cold and wet conditions, Kingdon-Ward was in his element, collecting all the seed he could reach. Among his new finds from here were *R. auritum*, a tender species with pale yellow to cream flowers growing from a huge boulder standing in a sandy bay, the bristly-leaved *R. exasperatum* and *R. venator* with 'fine scarlet flowers'. And on a high ridge above the monastery he found the beautiful dwarf species *R. pemakoense*, which has funnel-shaped, rose-purple flowers.

Below the monastery Kingdon-Ward expected to find the 165ft falls of legend, but he was disappointed to discover that they were, in fact, only 30–50ft. As the gorge plunged ever more steeply, the trail became obscured and progress was slowed – paths were hacked through thick undergrowth, sheer cliffs were scaled on narrow ladders and unsteady rope-bridges crossed deep ravines. Kingdon-Ward, who had a terrible head for heights, spent the time bracing himself for the crossings by combing the immediate area for more seed. In this way he gathered *R. taggianum*, which has lovely white fragrant flowers.

As they continued their descent, Kingdon-Ward confirmed the course of the river and finally dispelled the myth of giant waterfalls. Instead, he discovered that the river dropped over a series of smaller cascades. The party turned away from the Kongbo Tsangpo on 18 December and returned to Tumbatse along the Po-Tsangpo river gorge. This was full of big-leaved rhododendrons, magnolias, cherries and blue pines, and Kingdon-Ward found *Rhododendron montroseanum*, with pink flowers and big leaves similar to those of *R. sinogra: nde*. Arriving on 26 December, they gathered their belongings and began the trek back to Assam. Kingdon-Ward was very satisfied: not only had he collected many new plants, including *Berberis calliantha*, *Primula cawdoriana* and *Rhododendron cinnabarinum* ssp. *xanthocodon*, but he had solved the mystery of the Tsangpo gorge. His only regret was perhaps the season:

> What then, must it be like in the spring, when forest a mile long and a mile high would be in full bloom, thousands upon thousands of trees in flower together, making huge daubs, mounds and bands of colour in the dark gorge! … How they would light up the shadows! It is a spectacle which some future botanical explorer may enjoy … No collecting 'blind' for him! No rhododendron unseen!

Several more visits to north Burma followed. The 1926 expedition to the Seingku river was sponsored by a group of wealthy gardeners including Lionel de Rothschild, the rhododendron enthusiast from Exbury, near Southampton. The following year Kingdon-Ward returned to the Mishmi Hills and Nagaland in Assam and climbed Japvo Peak (9,890ft), which he had first climbed in 1918. Half-way up his 'guides' and porters asked him where the next water lay. At this point Kingdon-Ward realized that they were following him and not the other way round! Although he had a hopeless sense of direction, he used his good memory and years of travel experience and two days later they successfully made the summit, where 'I saw the first specimens of the magnificent *R. macabeanum* discovered by Sir George Watt. It reaches tremendous size, growing over forty feet high, with leaves fifteen inches long, and almost entirely covers the summit of Japoo[*sic*].' He made the first introduction of this species, which 'bears huge trusses of lemon-yellow flowers, stained plum-juice purple at the base'.

On his 1929 trip Kingdon-Ward moved east into Laos, and the next year, with the 4th Earl of Cranbrook as his companion, he visited the Ajung valley in north Burma. Now funded by the Royal Horticultural Society, he returned to Tibet in 1933. Two years later he was back again. Leaving Tezpur on the Brahmaputra River in Assam, he moved north through the hills towards Tibet. At Shergoan, near the Bhutanese border, he applied to the Governor of Tsona (now Cona), the first town inside Tibet, for permission to enter the country. Confident of a favourable reply, he went ahead to the scruffy hamlet of Lugathang, where he received an answer from the Governor. However, the message was written in Tibetan, and since neither he nor his servant could read the language they decided to continue to the larger town of Karta in order to find an interpreter. When they arrived Kingdon-Ward discovered that the letter was missing, lost among the baggage. Not one to worry too much about formalities, and reasoning that had entry been refused he would have been stopped by now, he carried on east to Chayul (now Qayu), visiting the magnificent monasteries of Sanga Choling (now Sangngagqoling) before arriving at Liliung on 18 July. Ten years earlier, when he had first stayed there, he had admired a great range of peaks to the north and it was these that he now determined to explore.

At Tumbatse he was welcomed by his old friends, and at Tongkyuk Dzong he was delayed by a Lassa police chief, Colonel Yuri. Kingdon-Ward had to talk fast to convince the Colonel that he was not a Bolshevik spy. Eventually his innocence was accepted and he was allowed to continue north, crossing the remote Sobhe La Pass and descending towards Yigrong (now Yi'ng) on the

Po-Yigrong river (now Yi'ong Nyangbo). Here, to his acute embarrassment, he was mistaken for the Colonel and greeted with the cry, 'Welcome, great Chief, welcome' and accorded a princely reception.

Kingdon-Ward was gripped by the desire to follow the Po-Yigrong river to its source in the mountains. He found that ascending the narrow, steep river gorge was almost as arduous as his descent of the Tsangpo, but the beautiful hemlock (*Tsuga* species) forests, full of birds and *Rhododendron niphargum*, kept him in good spirits. Above the village of Ragoonka, however, was a sheer granite cliff which had to be traversed along a tiny ledge. Kingdon-Ward was frozen by his dread of heights – just looking over the edge 'made me giddy, and I shrank back with that awful stab in the pit of the stomach which sudden fear can induce … My heart was in my mouth; the sight gave me physical pain.' As helpless as a small child, he had to be coaxed across by his bemused porters with the help of ropes.

On 20 August, after eighteen days, Kingdon-Ward crossed alpine meadows that were full of 'millions of flowers' of *Primula sikkimensis*, *Meconopsis* species and alliums to reach the glacier at the source of the river. It was a lowering, overcast day and cloud choked the valley, but on the pass above the glacier Kingdon-Ward turned to look back one last time:

> Suddenly the amorphous cloud thickened and curled; and the veil was rent, and there was revealed the most wonderful view of the actual source of the Po-Yignong my wildest fancy could ever have painted. If I had waited and dreamed for ten years for that brief glimpse only, I had not lived in vain. It epitomized a life's ambition; a worthwhile discovery in Asia, truly finished.

Contented, Kingdon-Ward now began the wearisome journey home, arriving back at Tezpur in mid-October. He had made a round trip of 1,000 miles and he had gathered hundreds of plant species. It was when he was unpacking his collection that the 'lost' and forgotten letter came to light. When it was translated, it turned out to be a refusal of permission to enter Tibet!

Kingdon-Ward undertook three more journeys in the late 1930s, to Upper Burma, the Burma–Tibet frontier and Assam. On his visit to Burma in 1937 he had another close encounter with death when he fell on a muddy and slippery mountain path and, unable to grab hold of anything to check his progress, slid down the mountainside. Just as he was about to disappear over a cliff he was brought to an abrupt halt by a painful but lifesaving bamboo spike which impaled his armpit.

When war was declared in 1939 Kingdon-Ward joined up. He received his previous rank of Captain and was attached to the Special Operations Executive. He was sent on missions to establish safe military corridors through Burma, avoiding the Japanese, and towards the end of the war he taught jungle survival to airmen at the School of Jungle Warfare in Poona, India. When the war ended, the Government of the United States employed him to locate aircraft that had been lost on the 'Hump', the mountainous area between India and China. Most of the 200–300 downed aircraft were casualties of the appalling weather rather than of enemy fire. It was on one of these missions that he first found the Manipur lily.

In 1948 he returned to the remote state of Manipur, sandwiched between Assam, Nagaland and north-west Burma, to collect on behalf of the New York Botanic Garden. He and Florinda had divorced ten years earlier after fourteen years of marriage, and in 1947 he had married a much younger second wife, Jean Macklin. On this, their first of six trips together, they gathered many bulbs of the lovely pink Manipur lily (which Kingdon-Ward now named *Lilium mackliniae* after Jean) on the mountainsides of Sirhoi Kashong:

> The half-nodding bell was a delicate shell-pink outside, like dawn in June, with the sheen of watered silk; inside, it was like faintly flushed alabaster … We sang, we shouted with joy; then moved on, up the hill, up the hill, up the hill, till we found ourselves surrounded by pink lilies … When a breeze swept through the meadow, hundreds of lilies bowed their heads and swung their bells to and fro, the whole slope twinkling and dancing joyfully.

In 1952 Jean wrote her first book, *My Hill So Strong*. In it she recounts their journey to the Lohit Valley in Burma, where they experienced a massive earthquake. Running from their tents they were thrown to the ground, and as Frank wrote in the *Statesman* (of India):

Kingdon-Ward was accompanied by his wife on several of his later expeditions. The photograph shows his second wife, Jean, in Lohit Valley. Note the state of her shoes!

We knew that we were helpless, and the surprising thing is that we talked so calmly to each other, frightened out of our wits as we were. I had the feeling that we were lying on a thin cake of rock crust which separated us from the boiling interior of the earth and that this crust was about to break up like an ice-flow in spring, hurling us to a horrible death.

The Kingdon-Wards survived the earthquake, having heeded the tremors of the preceding weeks and sensibly pitched their tents in open ground. The epicentre of the quake was just 25 miles away and it was one of the most powerful ever recorded, wreaking havoc throughout Burma. The rice terraces of the valley collapsed into a sea of mud, engulfing towns and villages, and a thick dust cloud blotted out the sun for days.

In 1956, Frank, now an energetic seventy-one years old, scaled Mt Victoria (10,016ft) in west central Burma. He followed this with a journey to Sri Lanka in 1956–7. He was planning yet another trip when he fell ill, lapsed into a coma and died on 8 April 1958 in London, aged seventy-three.

Without the advantage of the manpower used by his contemporaries elsewhere in China, Frank Kingdon-Ward was a resourceful solo plant hunter. His hard work was recognized in his lifetime by the Royal Horticultural Society, which awarded him the Victoria Medal of Honour in Horticulture in 1932 and the Veitch Memorial Medal in 1933. His friends in the Massachusetts Horticultural Society awarded him the George Robert White Memorial Medal in 1934, and eighteen years later he was nationally honoured for services to horticulture with the Order of the British Empire. However, it is perhaps the plants themselves, which include over 100 new rhododendron species, that most fittingly bear testament to the man. As Dr N.L. Bor aptly put it in his Introduction to *Pilgrimage for Plants*:

> those who are prepared to accept the penalties of one-man exploration – the physical hardship, the utter loneliness of months in a strange land among strange people, the nauseating dullness of a diet of *tsampa* [flour ground from roasted barley grains] washed down with rancid butter tea and all the inconveniences of travelling 'light' – must possess exceptional courage, determination, loyalty to their sponsors, and devotion to their purpose. These qualities, combined with modesty – for all the great explorers were modest men – are qualities which make men great, and Kingdon-Ward possessed them in full measure.

Frank Kingdon-Ward's Plant Introductions

The date beside each plant name is the date of its introduction into Britain.

Rhododendron wardii (1913)
Rhododendron = (Gk) *rhodo*, rose; *dendron*, tree
wardii = after Frank Kingdon-Ward

Beautiful saucer-shaped, clear yellow flowers, freely produced in May to June. This compact evergreen species has rounded deep green leaves and is influential in the breeding of yellow hybrids. Hybrids include 'Cowslip' (× *R. williamsianum*, pre-1937, primrose), 'Goldkrone' (× *R.* 'Alice Street', 1983, golden yellow), 'Hotei' (*R.* 'Goldsworth Orange' × [*R. soulei* × *R. wardii*], 1968, deep yellow), and Flava Group (× *R. yakushimanum*, pale yellow).

Ranges from Yunnan and Sichuan in China to south-east Tibet, growing on rocky hillsides, in scrub, conifer or mixed forests, and even limestone cliffs and swamps at 8,800–11,500ft (2,700–3,400m), reaching from 3–26ft (90cm–8m) tall.

Primula burmanica (1914)
Primula = from Lat. *primus*, first, i.e. early flowering
burmanica = (Lat.) from Burma

Gorgeous whorls of yellow-eyed, magenta-pink flowers are borne on 24in (60cm) stems in July to August. This superb candelabra primula has rosettes of coarse 12in (30cm) leaves. Kingdon-Ward also introduced *P. chungensis* (1913, pale orange flowers).

From north-east Burma and north-west Yunnan in China, growing in wet meadows and conifer forests at 8,800–10,500ft (2,700–3,200m).

Meconopsis betonicifolia (1924), syn. M. baileyi
Meconopsis = (Gk) poppy-like
betonicifolia = (Lat.) leaves like betony (*Statice officinalis*)

Delightful, 2in (5cm) wide, sky-blue, papery flowers with bright yellow anthers hover tantalizingly above rosettes of handsome hairy leaves on slender-leafed stems up to 4ft (1.2m) tall. A highly desirable perennial. The hybrid *M. × sheldonii* (× *grandis*, 1937) has produced some exceptional forms, e.g. 'Branklyn', a magnificent giant with 6ft (1.8m) stems, and 'Slieve Donard', with longer, pointed petals.

Ranges from north-west Yunnan in China to north-east Burma and south-east Tibet, growing in woods and by alpine meadow streams at 10,000–13,000ft (3,000–4,000m).

Primula florindae (1924)
Primula = from Lat. *primus*, first, i.e. early flowering
florindae = after Kingdon-Ward's first wife, Florinda

Deliciously scented, pendant, lemon yellow waxy bells are suspended on 24in (60cm) tall white-powdered stems from June to August. A wonderful, large species with clumps of luxuriant rounded deep green leaves up to 8in (20cm) long and 6in (15cm) wide. Kingdon-Ward also introduced *P. alpicola* (c. 1940) from south-east Tibet, with lovely pendant bells of purple or white.

From the Tsangpo basin, south-east Tibet, growing in spreading drifts in marshy places by streams at 11,000ft (4,000m).

Cotoneaster conspicuus (1925)
Cotoneaster = (Gk) *cotonem*, quince; *aster*, incomplete resemblance
conspicuus = (Lat.) noticeable

Abundant, glossy, bright red fruits crowd elegant arching branches in autumn. This splendid, medium-sized, spreading evergreen shrub bears a mass of white flowers in May to June among slender bright green leaves. An excellent form is 'Decorum', which is lower growing. Kingdon-Ward also introduced *C. sternianus* (1913), which has silver undersides to the leaves, pink flowers and bright orange-red fruits.

Occurs in south-east Tibet, growing on rocky hillsides and over boulders at 8,200ft (2,500m), reaching 6½ft (2m).

Rhododendron macabeanum (1928)

Rhododendron = Gk *rhodo*, rose; *dendron*, tree
macabeanum = after Mr M'Cabe, Deputy Commissioner for the Naga Hills

Resplendent, 12in (30cm) long, heavily veined shining dark green leaves, softly felted grey-white beneath, are topped by candelabra-like trusses of clear yellow, bell-shaped flowers in March to April. This is a quite magnificent large shrub or small tree, with silvery-white young growth emerging from bright scarlet buds. 'Mrs George Huthnance' (parentage unknown, 1981) is a New Zealand hybrid, with pink buds and primrose-yellow flowers.

Found only on hilltops in Manipur and Nagaland, north-east India, growing in birch forests or as pure stands at 7,800–8,800ft (2,400–2,700m), reaching 45ft (14m) tall.

Lilium mackliniae (1948)

Lilium = (Lat.) lily
mackliniae = after Kingdon-Ward's second wife, Jean Macklin

Charming, pale purple-rose, bell-shaped flowers appear on 8–36in (20–90cm) leafy stems in June. A choice and distinct lily.

Occurs only on Sirhoi Kashong near Urkhul in Manipur, India, growing in grassy mountain slopes and rocky places at 7,000–8,500ft (2,150–2,600m).

Contoneaster conspicuus, from Tibet, was one of Kingdon-Ward's finest shrubby introductions, with its elegant, arching branches and striking red berries.

BIBLIOGRAPHY

BOOKS

Allen, M., *The Tradescants*, Michael Joseph, London, 1964.

Allen, M., *The Hookers of Kew*, Michael Joseph, London, 1967.

Allen, M., *Plants That Changed Our Gardens*, David & Charles, Newton Abbot, 1974.

Amherst, A., *A History of Gardening in England*, John Murray, London, 3rd edn, 1910.

Banks, R.E.R., Elliott, B., King-Hele, D., Hawkes, J.G., Lucas, G.L.L. (eds.), *Sir Joseph Banks: A Global Perspective*, Royal Botanical Gardens Kew, London, 1994.

Beaglehole, J.C. (ed.), *The Endeavour Journal of Joseph Banks 1768–1771, Volumes I & II*, The Trustees of the Public Library of New South Wales in association with Angus & Robertson, Sydney, 1962.

Bisgrove, R.J., *The National Trust Book of the English Garden*, Viking, London, 1990.

Bower, F.O., *Joseph Dalton Hooker*, SPCK, London, 1919.

Briggs, R.W., *'Chinese' Wilson*, HMSO, London, 1993.

Brockway, L.H., *Science and Colonial Expansion: the Role of the British Royal Botanic Gardens*, Academic Press, London, 1979.

Cameron, H.C., *Sir Joseph Banks*, Angus & Robertson, Sydney, 1952.

Carter, H.B., *Sir Joseph Banks*, British Museum (Natural History), London, 1988.

Coats, A.M., *Flowers and Their Histories*, Hulton Press, London, 1956.

Coats, A.M., *The Quest for Plants*, Studio Press, London, 1969.

Coates, A., *Garden Shrubs and Their Histories*, Vista Books, London, 1963.

Cox, E.M.H., *Plant Hunting in China*, Collins, London, 1945.

Cox, P.A., *The Larger Rhododendron Species*, Batsford, London, 1990.

Cox, P.A., *The Smaller Rhododendrons*, Batsford, London, reprint 1990.

Cox, P.A. and Kenneth, N.E., *Encyclopedia of Rhododendron Hybrids*, Batsford, London, 1990.

Davies, J., *Douglas of the Forests*, Paul Harris Publishing, Edinburgh, 1980.

Douglas, D., *Journal kept by David Douglas during his travels in North America 1824–1827*, William Wesley & Son, London, 1914.

Elliott, B., *The Country House Garden*, Mitchell Beazley, London, 1995.

Elliott, B., *Victorian Gardens*, Batsford, London, 1986.

Fisher, J., *The Origins of Garden Plants*, Constable, London, 1982.

Flemming, L. and Gore, A., *The English Garden*, Michael Joseph, London, 1979.

Fletcher, H.R., *The Story of the Royal Horticultural Society*, Oxford University Press, Oxford, 1969.

The Scottish Rock Garden Club, *George Forrest, V.H.M. Explorer and Botanist*, Edinburgh, 1935.

Forrest, G., *Field Notes of Trees, Shrubs and Plants other than Rhododendrons Collected in Western China by Mr George Forrest 1917–1919*, Royal Horticultural Society, London, 1929.

Fortune, R., *Three Years' Wanderings in the Northern Provinces of China*, John Murray, London, 1847.

Fortune, R., *A Journey to the Tea Countries of China*, John Murray, London, 1852.

Fortune, R., *A Residence among the Chinese*, John Murray, London, 1857.

Fortune, R., *Yedo and Peking*, John Murray, London, 1862.

Hadfield, M., *A History of British Gardening*, Spring Books, London, 1969.

Hadfield, M., *Landscape with Trees*, Country Life, London, 1967.

Harvey, A.G., *Douglas of the Fir*, Harvard University Press, USA, 1947.

Harvey, J., *Early Gardening Catalogues*, Phillimore, London, 1984.

Healey, B.J., *The Plant Hunters*, Charles Scribner's Sons, New York, 1975.

Hepper, F.N., *Plant Hunting for Kew*, HMSO, London, 1989.

The Hillier Manual of Trees and Shrubs, David & Charles, Newton Abbot, 6th edn, 1996.

Hooker, J.D., *Himalayan Journals*, Ward, Lock, Bowden & Co, London, 2nd edn, 1891.

Hooker, J.D., *Rhododendrons of the Sikkim–Himalaya*, Reeve, Benham, Reeve, London, 1849

Hoyles, M., *The Story of Gardening*, Journeyman Press, London, 1991.

Huxley, L., *Life & Letters of Sir Joseph Dalton Hooker, Vol I*, John Murray, London, 1918.

Hyams, E. and MacQuitty, W., *Great Botanical Gardens of the World*, Thomas Nelson & Sons, London, 1969.

Jellicoe, G., Jellicoe, S., Goode, P. and Lancaster, M. (eds.), *The Oxford Companion to Gardens*, Oxford University Press, Oxford, 1986.

Kingdon-Ward, F., *Land of the Blue Poppy*, Cambridge University Press, Cambridge, 1913.

Kingdon-Ward, F., *The Mystery Rivers of Tibet*, Seeley Service, London, 1923.

Kingdon-Ward, F., *Riddle of the Tsangpo Gorges*, Edward Arnold, London, 1926.

Kingdon-Ward, F., *Plant Hunter's Paradise*, Jonathan Cape, London, 1937.

Kingdon-Ward, F., *Assam Adventure*, Jonathan Cape, London, 1941.

Kingdon-Ward, F., *Pilgrimage for Plants*, George G. Harrap & Co., London, 1960.

Leith-Ross, P., *The John Tradescants*, Peter Owen, London, 1984.

Lemmon, K., *The Golden Age of Plant Hunters*, J. M. Dent, London, 1968.

Loudon, J.C., *An Encyclopedia of Gardening*, Longman, Hurst, Rees, Orme & Brown, London, 1822.

Lyte, C., *Sir Joseph Banks: Eighteenth Century Explorer, Botanist and Entrepreneur*, David & Charles, Newton Abbot, 1980.

Lyte, C., *The Plant Hunters*, Orbis Publishing, London, 1983.

Lyte, C., *Frank Kingdon-Ward: The Last of the Great Plant Hunters*, John Murray, London, 1989.

Macqueen Cowan, J. (ed.), *The Journeys and Plant Introductions of George Forrest*, Oxford University Press, Oxford, for the Royal Horticultural Society, 1952.

Masson, F., *An Account of Three Journeys from Cape Town in to the Southern Parts of Africa, undertaken for the Discovery of New Plants towards the improvement of the Royal Botanical Gardens at Kew*, Addressed to Sir John Pringle, Bart., FRS, Royal Botanical Gardens Kew, 1775.

Morgan, J. and Richards, A., *A Paradise out of a Common Field*, Century, London, 1990.

Morwood, W., *Traveller in a Vanished Landscape – The Life and Times of David Douglas*, Gentry Books, London, 1973.

Mitchell, A., *Alan Mitchell's Trees of Britain*, HarperCollins, London, 1996.

Musgrave, J.T.T., *Innovation and the Development of the British Garden Between 1919 and 1939*, doctoral thesis, University of Reading, 1996.

O'Brian, P., *Joseph Banks, A Life*, Collins Harvill, London, 1987.

Ottewill, D., *The Edwardian Garden*, Yale University Press, Yale, 1989.

Philips, R. and Rix, M., *Bulbs*, Pan, London, 1989.

Philips, R. and Rix, M., *Perennials. Volume 1 Early Perennials*, Pan, London, 1991.

Philips, R. and Rix, M., *Perennials. Volume 2 Perennials*, Pan, London, 1991.

Philips, R. and Rix, M., *Shrubs*, Macmillan, London, 1994.

Recht, C. and Wetterwald, M.F., *Bamboos*, Batsford, London, 1992.

Rushforth, K., *Conifers*, Helm, Kent, 1987.

Scourse, N., *The Victorians and Their Flowers*, Croom Helm, London, 1983.

Spongberg, S.A., *A Reunion of Trees*, Harvard University Press, Harvard, 1990.

Stuart, D., *The Garden Triumphant*, Harper & Row, New York, 1988.

Thacker, C., *The History of Gardens*, Croom Helm, London, 1979.

Turrill, W.B., *The Royal Botanic Gardens, Kew*, H. Jenkins, London, 1959.

Turrill, W.B., *Joseph Dalton Hooker: Botanist, Explorer and Administrator*, Scientific Book Club, London, 1963.

Veitch, J.H., *Hortus Veitchii*, James Veitch & Sons Ltd, London, 1906.

Wilson, E.H., *Plant Hunting*, Harvard, Boston, 1927.

ARTICLES FROM PERIODICALS

Cornish Garden, No. 23, 1980, pp. 11–15, 'Concerning William Lobb'; Graham, B.

Cornish Garden, No. 24, 1981, pp. 13–21, 'Thomas Lobb 1817-1894'; Graham, B.

Cornish Garden, No. 25, 1982, pp. 17–18, 'Thomas Lobb and his "Japanese" collection'; Heriz-Smith, S.

Cornish Garden, No. 30, 1987, pp. 44–50, 'A short history of the lives and work of the brothers, William and Thomas Lobb'; Heriz-Smith, S.

Cornish Garden, No. 38, 1995, pp. 44–9, 'William Lobb (1809–1864) and his South American collections (1840-44)'; Heriz-Smith, S.

Cornish Garden, No. 39, 1996, pp. 55–7, 'William Lobb's collections (continued) from Peru, Ecuador, Southern Colombia and Panama (1842-44)'; Heriz-Smith, S.

The Gardeners' Chronicle, Vol. 83, p. 450, 'Mr F. Kingdon-Ward's tenth expedition in Asia'.

History Around the Fal, Part 5, pp. 55–77, 'A study of the Cornish plant hunters William and Thomas Lobb'.

Journal of the Horticultural Society, Vol. 1, pp. 208–24, 'Sketch of a visit to China in search of new plants'.

Journal of the Royal Horticultural Society, Vol. 68, pp. 161–9, 'Robert Fortune'.

Journal of the Royal Horticultural Society, Vol. 66, pp. 121–8, 153–64, 'David Douglas'.

Journal of the Royal Horticultural Society, Vol. 67, pp. 48–51, 'William and Thomas Lobb: two Cornish plant collectors'.

Journal of the Royal Horticultural Society, Vol. 72, pp. 33–5, 'A further note on the brothers William and Thomas Lobb'.

Journal of the Royal Horticultural Society, Vol. 80, pp. 265–80, 'Plant collectors employed by the RHS 1804–1846'.

Journal of the Royal Horticultural Society, Vol. 97, pp. 401–9, 'Robert Fortune and the cultivation of tea and other Chinese plants in the United States'.

ILLUSTRATION ACKNOWLEDGEMENTS

Jacket photograph and endpapers: **Images Colour Library**.

Ardea London/Jean-Paul Ferrero 104, 162–3; **Photographic Archives of the Arnold Arboretum** 148, 155, 165(t) (photographer: Ernest Henry Wilson), 165(b) (photographer: Ernest Henry Wilson); **Axiom Photographic Agency/Jim Holmes** 2–3, 182–3, 204–5; **Chris Gardner** 117, 121, 137; **Glenbow Collection** 65 ('Portage in Hoarfrost River', 1933, Sir George Back, Glenbow Collection, Calgary, Canada); **John Glover** 25, 53, 128, 159, 176, 194; **Jerry Harpur** 63; **Friedrich van Hörsten/Images of Africa Photobank** 38; **Anne Hyde** 56 (RSPB, Sandy Lodge, Beds); **Anthony Kersting** 22–3; **Andrew Lawson** 6, 19, 37, 46, 54, 84, 102, 130, 139, 154, 167, 170, 191, 192, 198, 209; By permission of the **Linnean Society of London** 39, 49, 55; **Linnean Society of London/Bridgeman Art Library, London/New York** 89 (from *The Rhododendrons of the Sikkim–Himalaya* by J.D. Hooker, 1849); **Toby Musgrave** 93; **National Library of Australia, Canberra, Australia/Bridgeman Art Library, London/New York** 27 (from the *Illustrated Sydney News Supplement*, Dec 1865 (wood engraving) by Samuel Calvert); **The Natural History Museum, London** 29; By courtesy of **The National Portrait Gallery, London** 13, 32; **Clive Nichols** 45, 51, 77, 122, 153, 217; **John Noble/Wilderness Photography** 70–1, 78; **Provincial Archives, Victoria, British Columbia, Canada** 61; **Royal Botanical Gardens, Kew, London/Bridgeman Art Library, London/New York** 83 (from *The Rhododendrons of the Sikkim–Himalaya* by J.D. Hooker, 1849); **Royal Geographical Society, London** © 180(t), 190, 199, 210(t), 210(b), 214; **Royal Horticultural Society, Lindley Library** 105, 177, 180(b); **Stapleton Collection/Bridgeman Art Library, London/New York** 144 (Sydenham: Crystal Palace, the Egyptian room, Tropical Plants, by Philip H. Delamotte, 1854); **TRIP/Eric Smith** 12; **Tropix/M. Auckland** 142.

INDEX

Page numbers in *italic* refer to illustrations

Abies
  A. amabilis 69
  A. grandis 54, 55, 69, 76
  A. procera 77
Aborigines 30
Acer griseum 174
Actinidia chinensis 158
Aerides rosea 152
agriculture 132
Aiton, William 33
Alcock, Sir Rutherford 126
Algoa Bay 44
Aloe dichotoma 47
Amaryllis
  A. belladonna 47, 53, 53
  A. disticha 41
American expeditions
  David Douglas 59
  William Lobb 141
'American Garden' 74–5, 99
Amoy 109
Anderson, Malcolm P. 200
Anemone hupehensis var. *japonica* 128,
  128
Araucaria araucana 56, 136–7,
  136–8, 139
Arbutus menziesii 67
Arnold Arboretum 164, 173
Arts and Craft movement 194–5
Assam 206, 212
Atuntze 203
Augusta, Dowager Princess of Wales
  33
Australia 26–30, 27

Bagshot Nursery 98–9
Banister, John 56
Banks, Sir Joseph 13, 13–36, 32
  at Kew 31, 33–5
  at Spring Grove 33
  childhood 14
  education 14
  herbarium 31

Labrador and Newfoundland survey
  15
  plant introductions 36
  see also Endeavour voyage
Banksia integrifolia 27, 29
Bartram, John 56
belladonna lily 47, 53, 53
Bentham, George 100
Berberis darwinii 151
bird-of-paradise flower 44, 45, 53
Blaauwberg 41
Blue Mountains 64, 66
Bokkeveld Mountains 46–7
bonsai 112–13
Borneo 150, 200
Botany Bay 27
Brown, Lancelot 'Capability' 34
Bryum argenteum 80
Buchan, Alexander 15, 18–19, 21
Bulley, Arthur Kilpin 179, 200
Burma 143, 145, 150, 211–12, 213,
  214
Bustard Bay 28

Calaveras Grove 148, 148–9
California 69, 72, 146–9
Camellia
  C. reticulata 197
  C. saluensis 176, 177, 197
Campbell, Archibald 82, 90, 94, 95–6
Canada 50
Canary Islands 48
Capel, Sir Henry 33
Cardiocrinum
  C. cathayanum 166
  C. giganteum 146
Cawdor, Lord 207
Ceanothus x *veitchianus* 151
Chengdu 166
Chile 136–7, 136–40
Chiloe Island 140
Chinese expedition maps
  Forrest 184

Fortune 108
  Kingdon-Ward 201
  Lobb, Thomas 141
  Wilson 157
Chiswick Gardens 58, 107
Chola Pass 94–5
Choongtam 92, 94
Chusan Archipelago 110, 112, 115,
  121
Clarkia pulchella 62
Clematis
  C. armandii 174
  C. montana var. *rubens* 154, 155
Cockqua 62, 68
Colorado 101
conifers in garden design 74, 75, 194
conservatories 144, 145–6, 194
Cook, Captain James 15, 16, 27, 39
Cook's Passage 30
Cornus kousa var. *chinensis* 167, 175
Cotoneaster conspicuus 208, 216–17,
  217
Cragside 194
Cryptomeria japonica 128
Crystal Palace 131, 144
Cunninghamia lanceolata 106, 112
Cupressus funebris 123

Dalhousie, Lord 81–2, 84, 96
Dali 182–3
Darwin, Charles 80, 97–8
David, Jean Pierre Armand 156
Davidia involucrata 155, 158, 159,
  174
Delavey, Jean Marie 156
Dicentra spectabilis 123
Dipelta floribunda 164, 174
Don, George 58
Donkia Pass 94
Douglas, David 55, 55–73
  at Glasgow Botanic Gardens 57
  at Scone Palace 57
  at Valleyfield 57

in California 69, 72
in Hawaii 72–3
in New England 58–9
Pacific North–west expedition *59*,
   59–69, *70–1*
plant introductions 76–7
Douglas fir *56*, 57, 76
Dowd, A.T. 147

Earl of Talbot 50
East India Company 123–4
Edwardian gardens 195
Ellena 160–1
Elliott, Brent 195
Embothrium coccineum 151
Emu 109
Endeavour voyage 15–31, *17*
   Java 30–1
   Madeira 16
   map *17*
   New Zealand 21–6, *22–3*
   reef damage 28, 30
   Rio de Janeiro 17–18
   St Helena 31
   Society Islands 21
   South Africa *31*
   Tahiti 19–20
   Terra Australis 26–30, *27*
   Tierra del Fuego 18–19
English Landscape School 74
Erebus 80
Erythronium grandiflorum 64

Falconer, Hugh 81
Farges, Paul Guillaume 160
fertilizers 132
Florilegium 29
flowering currant 62, *63*, 76
Foochow 118
Forbes, John 58
Forrest, George *177*, 177–95, *180*
   at Edinburgh Botanic Garden 178
   plant introductions 196–7
   Talifu uprising 191–2
   Tibetan uprising 181–8
   Yunnan exploration *184*, 191
Forsythia
   F. suspensa var. *fortunei* 129
   F. viridissima 115, 129
Fort Vancouver *61*

Fortune, Robert *105*, 105–27
   at Chiswick 107
   attacked by pirates 119–21
   Chinese expeditions *108*
      of 1853–6 124
      East India Company 123–4
      Horticultural Society 109–23
      United States Government 124
   education 107
   Japanese expedition 124–6
   *Journey to the Tea Countries of
      China* 125
   mugged by Chinese 114
   plant introductions 128–9
Fraser, John 56
Frederick, Prince of Wales 33

garden design
   Arts and Crafts influence 194–5
   conifers 74, 75, 194
   Edwardian 195
   Jekyll garden 195
   naturalistic plantings 100
   pinetum 75
   replica foreign scenery 100
   rhododendromania 99–100
   use of exotic plants 10
   see also theme gardens
'Gardenesque' 74
Garrya elliptica 76, 77
Gaultheria
   G. mucronata 19
   G. shallon 60
'General Sherman' 149
Gentiana sino-ornata 196
Great Barrier Reef 28
Great Exhibition 131
Green Island 28
guano 132
Gurney, Ned 72

Hall, George Rogers 126
handkerchief tree 155, 158, *159*, 174
Harvard 165, 166
Hauraki Gulf 26
Hawaii 72–3
Henry, Augustine 156
hippopotamus pits 43
Hoarfrost river *65*
Hodgson, Bryan 82

Hong Kong 109, 113
Hooker, Sir Joseph Dalton *79*,
   79–101, 143, 145
   at Kew 98, 100–1
   childhood 79–80
   in Colorado 101
   education 80
   Erebus voyage 80
   Flora Antarctica 82
   Flora of Ceylon 101
   Flora Indica 101
   Flora of New Zealand 97
   Genera Plantarum 100
   Himalayan Journals 97
   in Lebanon 101
   in Morocco 101
   plant introductions 102–3
   Rhododendrons of Sikkim–Himalaya
      98
   Sikkim expedition *81*, 81–97
Hooker, Sir William 57, 69, 79, 98
Horticultural Society 57–8, 73,
   109–23
Hottentots Holland Mountains 40–1
Hpimaw 206
Hwuy-chow 124
Hydrangea macrophylla 106
Hypericum hookeri 146

India 146
Islumbo Pass 90
Ixia viridis 41

Japan 124–6, 171–2
'Japanese Garden' 127
Jasminium nudiflorum 116, *117*,
   129
Java 30–1, 143, 149, 200
Jekyll, Gertrude 195
Jiangxi province 166
Jodrell Laboratory 101
Jurume 172

Kanchenjunga 91
Kangding 161, 164
Kanglanamo Pass 90
Karoo 42, 47
Kellogg, Dr Albert 147
Kerr, William 75, 106
Kerria japonica 106

Kettle Falls 64
Kew 10, 31, 33–5, 98, 100–1
Khasia Hills 146
Kingdon-Ward, Frank *199*, 199–215, *210*
  in Burma 211–12, 213, 214
  eastern Himalayas expedition 207 11
  in Laos 212
  plant introductions 216–17
  in Tibet 212–13
  war years 214
  in Yunnan 201–6
Kongbo Tsangpo monastery 208–9
Kyushu Island 172

Lagerstroemia indica 166
Lamorran 100
Lange Kloor 44
Laos 212
lawn mowers 132
Lebanon 101
Lee, Francis L. 126
Lepchas 85
liana cane bridge *180*
Lichiang Mountains *190*, 190–1
Lilium
  L. lancifolium 106
  L. mackliniae 214, 217
  L. pandicum 64
  L. regale 164, 169, *170*, 175
  L. wardii 208
Lingcham 90
Lithops species 46
Little Karoo 43
Litton, George 179, 190
Lobb, Thomas 97, 134, 141–6, 149–50
  plant introductions 151–2
Lobb, William 134, 136 41, 146–9
  plant introductions 152–3
Lonicera
  L. fragrantissima 121, 123
  L. standishii 123
Loudon, John Claudius 73–4, 133
Lupinus polyphyllus 62, 76
Lushan Hills 166
Lutyens, Sir Edwin 194–5
Lyons, Israel 14

Macklin, Jean *214*, 214

McLeod, A.R. 67
McLoughlin, John 60
McNabb, William 107
Madeira 16
Magnolia
  M. campbellii 82, 197
  M. sinensis 175
  M. virginiana 56
Mahonia
  M. aquifolium 61
  M. bealei 129
  M. japonica 123, 129
Manipur 214
manures 132
Maoris 24, 26
Masson, Francis *39*, 39–52
  in Canada and North America 50
  letter to Linnaeus *49*
  plant introductions 53
  in Portugal 48–50
  in South Africa 39–48, *42*, 50
  transatlantic circuit 48
Matthew, John 149
Meconopsis
  M. betonicifolia 198, 199, 208, 216
  M. integrifolia 161
  M. punicea 161
Menabilly 99
Menzies, Archibald 56, 146
Milford Sound *22–3*
Min valley 169
monkey puzzle tree 56, *136–7, 136–8, 139*
Monterey pine 69
Morocco 101
Morrow, James 124
Multnomah river 63
Myong valley 88

Nelson, David 34
Nepenthes 130
  N. rajah 149
  N. sanguinea 143, 152
New England 58–9
New Zealand 21–6, *22–3*
Niger 15
Ningpo *112*, 112–13
North American expeditions
  David Douglas *59*
  William Lobb *141*

Oaklands 99
Opium Wars 105, 106
orangeries 145
orchids 145
Oregon Mountains 147
Organ Mountains 140

Paeonia suffruticosa 106
Paoning 169
Paphiopedilum wardii 206
Parkinson, Sydney 15
Parkman, Francis 126
Parks, John 58
Parthenocissus henryana 168
Passiflora species 18
Pemakochung monastery 211
Pencarrow 134
Penstemon glaber 64
pesticides 132
Phalaenopsis amabilis 116, 152
Philippines 145, 150
Phormium tenax 24, 25
Piketberg 45
pinetum 75
Pinus
  P. bungeana 122, 123
  P. lambertiana 63
  P. longifolia 88
  P. ponderosa 64
  P. radiata 69, 77
Po-Yignong 213
Podocarpus species 26
Portugal 48–50
Potts, John 58
Primula
  P. alpicola 208
  P. bulleyana 196
  P. burmanica 206, 216
  P. capitata 93, *102*, 102
  P. florindae 208, *209*, 216
  P. pulverulenta 175
  P. sikkimensis 93, *102*–3
  P. vialii 191, 196
Protea
  P. cynaroides 46, *47*, 53
  P. grandiflora 43
Pseudotsuga menziesii 56, *57*, 76
Purangi Bay 24, 26
Purdom, William 167
Purea, Queen 20–1

Queen Charlotte's Sound 26

Reeves, John 106
regal lily 164, 169, *170*, 175–1
Resolution 39
Reynolds, John 15
rhododendromania 99–100
Rhododendron
    R. argenteum 82
    R. auritum 211
    R. campylocarpum 88
    R. catawbiense 56
    R. cinnabarinum 90, *93*, 93, 103
    R. dalhousiae 82, *83*
    R. edgeworthii 92
    R. exasperatum 211
    R. falconeri *84*, 84, 103
    R. fortunei 124, 129
    R. griersonianum 193, 197
    R. griffithianum 92, 100, 103
    R. haematodes 196
    R. hodgsonii 88, 103
    R. jasminiflorum 143
    R. kiusianum 172
    R. leucaspis 209, 211
    R. macabeanum 212, 217
    R. montroseanum 211
    R. niphargum 213
    R. nivecum 92–3
    R. obtusum 116
    R. occidentale 146
    R. pemakoense 211
    R. sinogrande 191, *192*, 196–7
    R. taggianum 211
    R. thomsonii 88, *89*, 103
    R. veitchianum 152
    R. venator 211
    R. wardii 203, 206, 216
Rhus typhina 9
Ribes sanguineum 62, *63*, 76
Richmond Lodge 33
Rio de Janeiro 17–18
Robinson, William 193, 194
Rockies 68
Roggeveld Mountains 45–6, 47
Ross, Captain James Clark 80

Sabine, Joseph 57–8
St Helena 31
Saldanha Bay 41

Salween 188–9
Santa Lucia Mountains 146
Sarawak *142*,145
Sargent, Charles 157, 164, 167, 168, 173
Scone Palace 57
Senecio cineraria 48
Sequoia sempervirens 72, 146–7, 149
Sequoiadendron gigantium 148–9, 152
Shanghai 113
Sibthorp, Humphrey 14
Sichuan province *162–3*
Sikkim 78, 81–97, 207
Singtam Soubah 93, 95
Snow Mountains 44
Society Islands 21
Solander, Daniel Carl 15, 18, 19, 24
Soochow 116
South Africa 31, 39–48, 50
Spathiglottis aurea 143
Spokane river 64
Sporing, Herman Didrich 15
Spring Grove 33
Stapelia incarnata 45
Strelitzia reginae 44, *45*, 53
sugar pine 63, 67
Sumatra 150
Swellendam 43

Tahiti *12*, 19–20
Talifu 191–2, 202
Tambur River *86*
Tayeto 21, 31
tea plant 123, 124
Tengyueh 201, 203
Terra Australis 26–30, *27*
Terror 80
theme gardens 74–5
    'American Garden' 74–5, 99
    'Gardenesque' 74
    'Japanese Garden' 127
    see also garden design
Thirsty Sound 28
Thompson, Dr Thomas 98
Thuja plicata 151–2
Thunberg, Carl Per 41–3, 45, 47
Tibet *210*, 212–13
    lama uprising 181–8
Tierra del Fuego 18–19
Tradescantia virginiana 10

Tradescent, John, the elder 9–10, 56
Tradescent, John, the younger 10
Trillium grandiflorum 50
Tropaeolum speciosum 151
Tsangpo river 207, 208, 211
Tumloong *96*
Tupaia 21, 24, 31

Umbellularia californica 67
Valleyfield 57

Vanda caerulea 97, 145, *153*, 153
Veitch, John 126, 127, 133, 136
Veitch nurseries *131*, 133–4, *135*
Veitch, Sir Harry 155, 156, 160
Viburnum
    V. davidii 174–5
    V. plicatum 115
    V. rhytidophyllum 168

Wa–Shan Mountains 161
wake robin 50
Wallanchoon 88
Ward, Harry Marshall 199
Watt, Sir George 212
Wedgwood, John 58
Weigela
    W. florida 115, 129
    W. rosea 115
Wellingtonia gigantea 148
Wells Williams, S. 124
Wilson, Ernest *155*, 155–73, *165*
    Chinese expeditions *157*, 157–71
    Japanese expeditions *157*, 171–2
    plant introductions 174–5
Wisteria sinensis 106

Xuanhan 169

Yakla Pass 94–5
Yunnan 181, *184*, *190*, 191, 201–6, *204–5*

Zantedeschia aethiopica 50, *51*, 52, 53
Zemu Samdong 92